T0320678

The Medical Almanac

Galen

The
Medical
Almanac

*A Calendar of Dates of Significance to the Profession of Medicine,
Including Fascinating Illustrations, Medical Milestones,
Dates of Birth and Death of Notable Physicians,
Brief Biographical Sketches, Quotations,
and Assorted Medical Curiosities and Trivia*

Pasquale Accardo, MD

 Humana Press • Totowa, New Jersey

© 1992 The Humana Press Inc.
999 Riverview Drive
Totowa, New Jersey 07512

Brief selections from this work appeared in *Missouri Medicine*, **83** (1986) in a section entitled *This Date in Medical History*.

Printed in the United States of America.   10 9 8 7 6 5 4 3 2 1

Library of Congress Cataloging in Publication Data
Main entry under title:

Accardo, Pasquale J.
      The medical almanac: a calandar of dates of significance to the profession of medicine, including fascinating illustrations, medical milestones, dates of birth and death of notable physicians, brief biographical sketches, quotations, assorted medical curiosities and trivia / Pasquale Accardo.
                  233 + *vii* pp.   15.24 x 22.86 cm.
      Includes bibliographical references.
      ISBN 0-89603-181-0
      I. Title.
      [DNLM: 1. Medicine. 2. Physicians —biography. WZ 112 A169m]
R104.A33 1991
610—dc20
DNLM/DLC for Library of Congress   91-20882 CIP

Preface

Now you Merely Acknowledge calendric resonances, the anniversary view of history, and catalogue them by Alphabetical Priority.
—John Barth, Letters

The history of medicine is a microcosm for the whole of human history. Its advances parallel progress in science and philosophy, civilization and ethics. Its pageantry mirrors all the triumphs and follies of human history. Osler commented that "the really notable years in the annals of medicine are not very numerous."[1] And indeed most calendars and almanacs record but few medical milestones.

The present collection has been made over a period of years. Its method is that of a commonplace book: "What have we to do but to set down this holy man's name in the calendar of saints?"[2] The persons herein represented include predominantly physicians, anatomists, and barber surgeons, with some few pharmacist-apothecaries and alchemists, an occasional chemist, biochemist, or physiologist, an infrequent nurse, and a rare medical school botany professor of *materia medica*. The milestones cover the dates of great discoveries, first (and often contested) operations, publications, and presentations. The feast days of holy patrons of those suffering various diseases are recalled, along with the dates of birth (and sometimes baptism), of medical graduation, and of the deaths of famous physicians. (It will be remembered that, with the exception of the feast of the Birth of Saint John the Baptist, saints are celebrated on the dates of their death—their birth into eternal life.) These entries are accompanied by brief biographies, anecdotes, and quotations when relevant. Eponyms are noted with the recognition that many of the physicians so honored often had a much deeper impact on the history of the profession. All specialties, all time periods, all national-

Table 1

Mark Akenside	Pio Baroja	Gottfried Benn
André Breton	Robert Bridges	Thomas Browne
Mikhail Bulgakov	Thomas Campion	Hans Carossa
Louis Céline	Anton Chekhov	George Crabbe
Michael Crichton	A. J. Cronin	Thomas Dooley
Arthur Conan Doyle	Georges Duhamel	Richard Austin Freeman
Oliver St. John Gogarty	Oliver Goldsmith	Johann Christian Gunther
Oliver Wendell Holmes	John Keats	Carlo Levi
Moses Maimonides	Bernard Mandeville	W. Somerset Maugham
Giovanni Meli	Silas Weir Mitchell	Axel Munthe
Max Nordau	Walker Percy	François Rabelais
James Ross	Johann Scheffler	Friedrich von Schiller
Arthur Schnitzler	Tobias Smollett	Eugène Sue
Frederick Treves	Henry Vaughan	William Carlos Williams
	John Wolcot	

ities are represented. Physicians' contributions to art, literature, law, philosophy, sport, cartography, geography, and exploration, inventions, physics, mineralogy, natural science, and biology are recognized. Epidemics, patients, and quacks provide some filler. The inclusion of some items may appear debatable, but can usually be defended.[3,4] It is of interest to note how many physicians (might have) had literary careers (Table 1) and how few excelled in the plastic arts (Table 2). Is there a neuropsychological explanation for this?[5]

This is mainly a browser's choice, with the only persistent criteria being medical importance and human interest (and a slight bias toward the editor's personal heroes).[6] I have attempted to steer a firm middle course between the narrow-minded hagiography of August Comte's calendar of the saints of positivism in his *Catéchisme positiviste* and the unpublished,

Table 2

Thomas Bartholin	Jerome Cardan	Robert Fludd
John Hill	Henry Hicks	Arthur Keith
Henry Pemberton	Hugh Williamson	Thomas Young

but widely known *Almanach antisuperstitieux* of Jean-Antoine-Nicolas Caritat, Marquis de Condorcet,[7] on the one hand, and the facile humor of Pudd'nhead Wilson's Calendar (Mark Twain) and *Armour's Almanac,*[8] on the other. *Crescit eundo;* the compiler welcomes corrections and additions to this essay.

What is reflected here is a particular view of the nature of medicine and the essence of true physicianhood. The "scientific hunters" make their discoveries on a particular date, usually early in their lives. The practitioners and teachers, long-lived "humanistic shamans," may not even be remembered by dates of birth and death, but only by a "floruit." The present history focuses more on individual human beings than on global trends or the evolutionary movement of civilizations. Although the profession combines aspects of both science and art, medicine is viewed predominantly as an art.[9-11] and the main artistic achievement of every physician is his or her own life.

Now let us praise famous physicians.

References

1. Osler W: Harvey and his discovery, In: McGovern JP, Roland CG eds. *The Collected Essays of Sir William Osler, vol. I, The Philosophical Essays*, The Classics of Medicine Library, Birmingham, 1985, p. 340.
2. Erasmus D: The apotheosis of Caprio, In: Jackson WTH, ed, *Essential Works of Erasmus*, Bantam Books, New York, 1965, p. 220.
3. Accardo PJ: Dante and Medicine: The Circle of Malpractice. *Southern Medical Journal, 1989;* **82:** 624–628.
4. Carter RB: *Descartes' Medical Philosophy: The Organic Solution to the Mind-Body Problem.* Baltimore: Johns Hopkins, 1983.
5. Accardo PJ, Haake C, Whitman BY: The learning disabled medical student. *Journal of Developmental and Behavioral Pediatrics, 1989;* **10:** 253–258.
6. Accardo PJ: William John Little (1810–1894) and cerebral palsy in the 19th century. *Journal of the History of Medicine and Allied Sciences, 1989;* **44:** 56–71.
7. Jaki SL: *The Road of Science and the Ways to God*, Chicago, University of Chicago Press, 1978, pp. 145–150.
8. Armour R: *Armour's Almanac*, McGraw-Hill, New York, 1962.
9. Maull N: The practical science of medicine. *J Med Philos 1981*, 6:165–182.
10. Munson R: Why medicine cannot be a science. *J Med Philos 1981*, 6:183–208.
11. Accardo PJ: *Diagnosis and Detection: The Medical Iconography of Sherlock Holmes.* Rutherford, New Jersey: Fairleigh Dickinson University Press, 1987.

To

Barbara Whitman, PhD

teacher
therapist
clinician
researcher
vocalist
musician
collaborator
friend

For having a sense of humor to appreciate the humor in
these assorted curiosities about "real" doctors.

January

January, month of empty pockets!...Let us endure this evil month, anxious as a theatrical producer's forehead.

—*Colette, Journey for Myself*

January 1

291 BC

Aesculapius was brought to Rome from Epidaurus at the insistence of the Sibylline Books after a plague in 293 BC. Legend told how the sacred snake itself chose the Insula Tiberina for its abode, and there on January 1, 291 BC, a temple was dedicated to *Aesculapius*.

Aesculapius

1788

Posthumous publication of *John Hunter*'s last work, *A Treatise on the Blood, Inflammation, and Gun-Shot Wounds.*

1820

On 1st January, 1820, a general convention consisting of delegates from different parts of the United States was held in Washington for the purpose of formulating a National Pharmacopoeia.

John Hunter

b.1849

Alois Epstein, a Bohemian pediatrician, first described Epstein's pearls—small white epithelial masses on the newborn palate.

1907

Pure Food and Drug Act goes into effect.

January 2

1968

Dr. Christiaan Barnard performed the first successful (19 months) human heart transplant on Cape Town dentist, *Philip Blaiberg.*

January 3

b.1840

Joseph de Veuster, a Belgian, joined the Society of the Sacred Hearts of Jesus and Mary (Picpus Fathers) as *Father Damien.* His work with lepers on Molokai permanently linked his name with that disease.

January 4

1885

Dr. William West Grant of Davenport, Iowa, diagnosed a perforated appendix in 22-year-old *Mary Gartside* and then completed the first successful appendectomy in America.

January 5

b.1826

Edme-Félix-Alfred Vulpian was a French physician and neurophysiologist who is now remembered eponymously by a scapulohumeral atrophy and a conjugate deviation in which the eyes turn to one side after an attack of apoplexy.

b.1874

Joseph Erlanger, American physician and neurophysiologist who was awarded the 1944 Nobel Prize for his work in analyzing the action potentials of nerve impulses.

d.1941

William Browning, a Brooklyn neuroanatomist who served at Long Island College Hospital. He described *Browning's vein,* the middle cerebral vein.

January 6

b.1714

Percival Pott, English surgeon of St. Bartholemew's Hospital. In 1756 he fell from his horse and fractured his ankle; in 1769 he published his *Some Few Remarks upon Fractures and Dislocations,* which included a description of a fracture of the lower part of the fibula and of the malleolus of the tibia with outward displacement of the foot—*Pott's fracture.* His *Remarks on that kind of Palsy of the lower limbs which is frequently found to accompany a curvature of the spine* (1779) described the kyphosis seen in tuberculous spondylitis (*Pott's disease*). In industrial medicine he discovered an occupational cancer in chimney sweeps. "The desire of health and ease, like that of money, seems to put all understandings, and all men, upon a level."

d.1878

William Stokes. This Irish physician specialized in chest diseases. He described *Cheyne–Stokes respiration* in which cycles of increasing depth and rate alternate with decreasing depth and apnea. Adams–Stokes syndrome or heart block is characterized by slow and occasionally irregular pulse, vertigo, syncope, and pseudoepileptic convulsions.

d.1981

A.J. Cronin, Scottish physician and novelist whose medical reminiscenses provided the substance for the radio and television series, *Dr. Finley's Casebook.*

January 7

1579

First medical course (Prima de Medicina) in the Americas starts at the Royal Pontifical University of Mexico, which had been founded in 1551 by a decree of *Charles V.*

1785

First cross-channel flight in a hydrogen balloon is accomplished by *Jean-Pierre Blanchard* and *Dr. John Jeffries.*

January 8

1800

Victor of Aveyron, the "Wild Boy" raised by animals, is captured.

b.1826

Paul Louis Duroziez. This Parisian physician first described a double (systolic and diastolic) intermittent murmur over the femoral arteries as a sign of aortic insufficiency. *Duroziez's disease* is congenital mitral stenosis.

d.1925

Guido Banti, an Italian physician and pathologist who was one of the founders of modern hematology. *Banti's disease* or splenic anemia is characterized by congestive splenomegaly, anemia, and leukopenia secondary to cirrhosis of the liver.

January 9

b.1778

Thomas Brown, Scottish physician, philosopher, and poet. He studied philosophy, medicine and law at the University of Edinburgh, practiced medicine briefly, and was one of the leaders of the "common sense" school of philosophy. He wrote commentaries on Hume and on Erasmus Darwins's *Zoonomia.*

b.1817

Nathan Smith Davis was born in a log cabin. He founded the American Medical Association to reform medical education and was the first editor of the *Journal of the American Medical Association.* His son and grandson each bore his name, and also pursued careers in medicine.

1929

The first clinical application of crude penicillin by **Alexander Fleming,** who treated an infected antrum by washing out the sinuses with dilute penicillin broth. The patient was his laboratory assistant, *Stuart Craddock.*

January 10

d.1654

Nicholas Culpeper, English physician, botanist, and astrologer. He wrote *Semeiotica Uranica, or an Astrological Judgement of Diseases* (1651) as well as a *Directory for Midwives.* In 1653 *The English Physician* was added to his famous *Complete Herbal.* He became a character in Kipling's *A Doctor of Medicine.*

d.1778

Karl von Linné (Carolus Linnaeus), Swedish botanist, naval physician, and taxonomist. His *System of Nature* (1735) introduced a classification utilizing binomial nomenclature. In 1745 he published one of the earliest descriptions of aphasia. He coined the names for more flora and fauna than any other person in history.

1836

Raj Kristo Dey becomes the first Hindu medical student to dissect a human body.

b.1866

Ludwig Aschoff. This German pathologist developed the concept of the reticuloendothelial system as a protective mechanism. *Aschoff's nodule* is an inflammatory body in heart muscle characteristic of a rheumatic process.

b.1887

Robinson Jeffers, American physician, forester, pessimistic poet. Among his works are *Tamar and other poems, Roan Stallion,* and *Medea.*

d.1879

Jacob Bigelow, American botanist and medical educational reformer. He wrote *Florula Bostoniensis* (1814); *Bigelowia Menziesii* is named for him. Of his attempts at verse *Osler* commented, "It is remarkable how many physicians write poetry, or what passes as such." The verse following recalls both *Ammian*'s poem in the Greek Anthology:

> *Nicias, a Doctor and musician,*
> *Lies under very foul suspicion.*
> *He sings, and without any shame*
> *He murders all the finest music:*
> *Does he prescribe? Our fate's the same,*
> *If he shall e'er find me or you sick.*

as well as Phyllis McGinley's assault on physicians practicing literary license in her "Complaint to the American Medical Association" in *A Pocketful of Wry.*

d.1942

Graham Steell, English physician whose entire name is eponymized in the early diastolic murmur of pulmonary valve insufficiency secondary to pulmonary hypertension as in mitral stenosis.

January 11

b.1638

Niels Stensen, Danish anatomist, geologist. A convert to Catholicism, he became a priest and eventually Bishop of Titiopolis. In his *Observationes anatomicae* (Leiden, 1662) he described the secretory duct of the parotid gland that is named after him. He also demonstrated that compression of an animal's aorta caused paralysis of the posterior portions of the body. He was actually born on New Year's day under the Julian calendar then still in use in Protestant countries.

d.1753

Sir Hans Sloane, English physician, naturalist, and botanist. He wrote a *Natural History of Jamaica,* and his museum and library formed the basis of the British Museum. He believed that all medical knowledge should be free and to that end bought and publicized a secret cure for rabies.

b.1814

James Paget, English physician, surgeon, and pathologist. There are two *Paget's* diseases: Osteitis deformans described in *A Form of Chronic Inflammation of Bones* (1877) and a precancerous affection of the areola and nipple described in *On Diseases of the Mammary Areola Preceding Cancer of the Mammary Gland* (1874).

James Paget

b.1842

William James, American physician, philosopher, and psychologist. His books include *Principles of Psychology* (1890), *Varieties of Religious Experience* (1902), and *Pragmatism* (1907). One of the founders of physiological psychiatry, he coformulated the *James–Lange theory* according to which emotion is the feeling aroused by bodily reactions; the latter are the cause and not the result of the emotion. *Dewey* described him as "a Columbus—an explorer of the inner world."

d.1792

William Brown, member of a prominent Maryland medical family. He wrote a 32-page pamphlet that was the first pharmacopeia published in the United States (1778). He succeeded Benjamin Rush as Surgeon General of the Middle Department during the War for Independence.

1922

Insulin is first administered to a 14-year-old boy afflicted with diabetes mellitus.

1964

Surgeon General's report on the harmfulness of smoking is published.

January 12

1725

A dated preface to the third edition of *Newton's Principia* thanks "*Henry Pemberton, MD,* a man of the greatest skill in these matters" for seeing this edition through the press. *Dr. Pemberton,* a physician and physicist, later published *A View of Sir Isaac Newton's Philosophy.*

b.1816

Ludwig Traube, German physician and pathologist. He introduced temperature curves and pioneered in the use of animals in medical experimentation. *Traube's semilunar space* is an anatomical area that is usually tympanitic because of the underlying stomach; the percussion note changes when there is pulmonary emphysema or pleural effusion. *Traube's double sound* is the tone over the femoral vessels in the presence of aortic and tricuspid insufficiency.

1861

The French otologist, *Prosper Ménière* (1799–1862) published his account of a new syndrome characterized by vertigo, tinnitus, and progressive deafness.

b.1902

Hiram Houston Merritt, American neurologist. With *Tracy Putnam,* he discovered the anticonvulsive, phenytoin (Dilantin).

January 13

1753

Oliver Goldsmith admitted to the medical school of the University of Edinburgh. "Had he not learned in suffering what he taught in song?" (*Osler*)

b.1867

Francis Townsend was an American physician who secured, in 1934, more than 25 million signatures in behalf of a proposal to pay $200 a month to everyone over 60 years of age. The money was to have been raised by means of a 2% business tax; his influence dwindled after the 1935 passage of the Social Security Program.

January 14

1794

The first successful Cesarean section is performed by *Dr. Jessee Bennett* on his wife *Elizabeth Hog Bennett* in Virginia.

b.1875

Albert Schweitzer, Franco-German musician, philosopher, theologian, and medical missionary. His books include *The Quest for the Historical Jesus* and *On the Edge of the Primeval Forest.* The key to his philosophy was *Ehrfurcht vor dem Leben,* "reverence for life." In 1913 he founded Lambarene Hospital in French Equitorial Africa; the inscription outside this jungle hospital read, "Here, at whatever hour you come, you will find light and help and human kindness." He was awarded the 1952 Nobel Peace Prize.

Albert Schweitzer

January 15

b.1836

Karl Rohr, German embryologist and gynecologist. *Rohr's stria* is a layer of the placenta.

b.1873

Rupert Waterhouse, English physician who described the Waterhouse–Friderichsen syndrome, a malignant form of meningococcal meningitis with adrenal hemorrhage and collapse, purpura, and shock.

January 16

b.1653

Johann Konrad Brunner, Swiss physician and anatomist. In his *Glandulae duodeni* (Frankfurt, 1687), he described the mucous glands in the submucous layer of the duodenum that are named for him. He also researched the effects of removing the pancreas from dogs.

d.1943

Sir William Arbuthnot Lane, English surgeon. He became a Fellow of the Royal College of Surgeons when only 26 years of age. He was interested in the human skeleton, its congenital deformities, and the extent to which it is prone to accident. He pioneered in the use of vanadium steel bars, plates, and screws in orthopedic surgery and devised an operation for brain decompression by the removal of skull segments. *Lane's kinks* are a series of variable flexures in the intestinal canal; *Lane's disease* is a combination of visceroptosis and autointoxication. Many patients were operated on for this mythical condition. *Dr. Lane* had great diagnostic skills and was one of **Conan Doyle's** medical models for *Sherlock Holmes;* he was rumored to be the original for *Doctor Cutler Walpole* in **Bernard Shaw's** *The Doctor's Dilemma.*

d.1945

John Kellogg, American surgeon. He worked at the Battle Creek Sanitarium in Michigan, wrote extensively on nutrition, diet, and digestion, and developed Corn Flakes for the company that today bears his name.

January 17

Saint Anthony of Egypt (c.251–356); St. Anthony's fire is variously erysipelas, bubonic plague, typhus, or yellow fever. St. Anthony's dance was ergotism or chorea. He was also the patron saint of the Order of Hospitalers of St. Anthony.

Saint Anthony of Egypt

1536

François Rabelais is absolved of apostasy by Pope Paul III and allowed to resume the practice of medicine in Montpellier. (This was part of an odd maneuver to turn him into a secular priest.)

1776

Drs. Daniel and Robert Perreau (twins) are executed at Tyburn for forgery.

b.1860

Anton Chekhov, Russian physician and writer. His plays include *The Sea Gull, Uncle Vanya, The Three Sisters*, and *The Cherry Orchard.* "I feel more confident and more satisfied with myself when I reflect that I have two professions and not one. Medicine is my lawful wife and literature my mistress. When I get tired of one I spend the night with the other. Though it's disorderly, it's not so dull, and besides, neither of them loses anything from my infidelity."

1912

Alexis Carrel starts a chicken cardiac cell culture. These embryonic fibroblasts will be kept alive until April 26, 1946.

d.1912

Joaquín Albarran y Dominguez, a Cuban-born French urologist. He was one of the founders of modern genitourinary surgery, an innovator in prostate surgery, and the inventor of a cystoscope used for urethral catheterization.

b.1927

Thomas Anthony Dooley, American medical missionary. "The cancer went no deeper than my flesh. There was no cancer in my spirit."

January 18

d.1719

Sir Samuel Garth, English poet and physician. His burlesque poem, *The Dispensary,* humorously recounts the efforts of the Royal College of Physicians to set up a dispensary for the London poor against the determined opposition of the apothecaries, who refused to supply medicines cheaply after the physicians gave their services for free.

> *Let them, but under their superiors, kill,*
> *When Doctors first have signed the bloody bill.*

He was the only medical member of the celebrated Kit Kat Club; *Pope* said that he was "the best good Christian without knowing it."

b.1779

Peter Mark Roget, English physician and thesaurus compiler.

January 19

1632

Dr. William Harvey is fired by St. Bartholomew's Hospital because his frequent attendance upon the king led him to be as frequently absent from the hospital; "therefore, *Dr. Andrewes,* physician in reversion...do supply the same place whereby the said poor may be more respected."

January 20

Feast of *Saint Sebastian*, patron of those suffering from epidemic pestilence or plague.

b.1580

Samuel Fuller, the doctor on the Mayflower. When he was loaned from Plymouth colony to Charlestown, he wrote back, "Many are sick, and many are dead, the Lord in mercy look upon them."

b.1840

Eugène Sue, French naval surgeon, socialist, and romantic novelist. His most famous work was *The Wandering Jew* (1844–1845). His father and grandfather were anatomists and surgeons.

January 21

d.1733

Bernard Mandeville, English physician and philosopher. He received his medical degree from Leiden in 1691. His *Enquiry into the Origin of Moral Virtue or the Fable of the Bees, or Private Vices, Public Benefits* (1714) declared virtue to be a delusion and identified selfishness as the cause of all human advances. His principal medical contribution was his *Treatise of the Hypochondria* (1711).

January 22

1973

United States Supreme Court decision permitting legal abortion.

b.1592

Pierre Gassendi, French philosopher and scientist; an atomist and a skeptic, an astronomer and a mathematician. His *Elegans de septo cordis pervio observatio* (1639) and *De foetus formatione* (1640) describe the foramen ovale.

b.1811

Edwin Hamilton Davis, American physician and archeologist. He recorded his work on mound builders in *Ancient Monuments of the Mississippi Valley* (1847), which became the first research monograph to be published by the Smithsonian Institution.

d.1818

Caspar Wistar, Philadelphia physician and anatomist. His *System of Anatomy* was the first American medical textbook (1811). In 1818 *Thomas Nuttall* named the beautiful plant *Wistaria* after him. (The spelling has always been in contention.) He is also remembered in the Wistar Institute and a well known line of white rats.

b.1855

Albert Ludwig Siegmund Neisser, German dermatologist and bacteriologist. He discovered the gonococcus in 1879. The genus *Neisseria*, which includes the organisms that cause gonorrhea and newborn ophthalmia, is named for him.

January 23

b.1835

William Henry Broadbent, London physician. *Broadbent's sign* is visible retraction of the chest wall in adherent pericardium. He also wrote on apoplexy and aphasia.

1849

The degree of Doctor of Medicine is conferred on *Elizabeth Blackwell*. "I was a young woman living at home with nothing to do in what authors call 'comfortable circumstances.' But I was wicked enough not to be comfortable. 'Why not be a nurse?' said the doctor. 'Because I prefer to earn a thousand rather than twenty pounds a year.'"

d.1854

Robert Montgomery Bird, American physician, playwright, novelist, and editor.

1953

Jonas Salk reports successful preliminary results for a new polio vaccine.

January 24

49 BC

Marcus Tullius Cicero concludes a letter to *Titus Pomponius* (*Atticus*): "If I was writing myself, this letter would have been longer, but I have dictated it because I am suffering from ophthalmia."

d.1848

Horace Wells, a pioneer in the history of anesthesia, becomes addicted to and unbalanced by chronic inhaling of nitrous oxide, ether, and chloroform, and commits suicide—the first to do so under anesthesia.

January 25

b.1813

James Marion Sims, pioneer American gynecologist. He discovered that a duckbill vaginal speculum—at first a bent spoonhandle—with the woman lying on her left side with her right thigh drawn up (*Sims's position*) permitted him to see "everything, as no man had ever seen before." He invented a surgical treatment for vesicovaginal fistula, was the first to suture with silver wires, and made the first diagnosis of a distended gall bladder—and performed the first cholecystotomy to drain it. He wrote *Clinical Notes on Uterine Surgery* (1866). When he had announced his career choice, the paternal response was far from encouraging: "It is a profession for which I have the utmost contempt. There is no science in it. There is no honor to be achieved in it, no reputation to be made, and to think that *my* son should be going around from house to house through this country with a box of pills in one hand and a squirt in the other, to ameliorate human suffering is a thought I never supposed I should have to contemplate."

b.1874

W. Somerset Maugham, physician and novelist. His works include the autobiographical *Of Human Bondage, The Moon and Sixpence, Cakes and Ale,* and *The Razor's Edge.* In *The Summing Up* he opined, "I do not know a better training for a writer than to spend some years in the medical profession."

January 26

Feast of *Saint Paula.* She was a disciple of *Saint Jerome* and was the first person to teach the art of nursing.

1841

Dr. D.J. West writes a letter to the *Lancet* with the first description of sudden massive spasms in infancy—what would later be called infantile spasms, hypsarrhythmia, or *West syndrome.* He was describing his own son.

1875

A patent is awarded to *Dr. George F. Green* of Kalamazoo, Michigan, for the electric dental drill.

b.1891

Wilder Penfield, a pioneer in epileptology and cerebral localization. "Don't always be a physician!"

January 27

1941

Penicillin first injected into a human subject; she had no infection and was made ill by a contaminant.

b.1621

Thomas Willis, English iatrochemical physician and anatomist. He is credited with the earliest descriptions of pertussis, typhoid fever, myasthenia gravis, and the 11th cranial nerve (the *nerve of Willis*). He was the first to distinguish diabetes mellitus from diabetes insipidus; he named puerperal fever and developed the concepts of lethargy and hysteria. He coined the term "neurology" and is probably best remembered in the eponym given to the arterial anastomoses at the base of the brain, the *circle of Willis.* The copper plates in

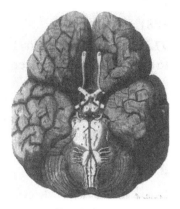

From: *Anatomy of the Brain and Nerves*

his classic *Cerebri Anatome* (1664) were the work of **Christopher Wren.**

January 28

d.1730

Antonio Vallisnieri, Italian physician, naturalist, entomologist, and geologist. He wrote treatises on the ostrich, the chameleon, the anatomy of worms, and the importance of the ovum. His name is remembered in the eponym for water weed, eelgrass or tape grass—*Vallisneria spiralis.*

1829

The "Resurrection man," **William Burke,** who was responsible for upwards of 30 to 40 murders and who gave a verb to the English language was hanged—and anatomized.

1949

John F. Enders, American bacteriologist, reports the culturing of poliovirus in *Science.* This step was critical to the later development of the Salk (1955) and Sabin (1957) vaccines. **Dr. Enders** later developed a vaccine for measles.

January 29

1881

Theodore Billroth (1820–1894) performs the first successful gastrectomy for stomach cancer on *Theresa Heller*, a 43-year-old mother of eight. Founder of the Vienna school of surgery, he had previously performed the first esophagectomy (1872), as well as the first laryngectomy (1873). He was also an accomplished musician and an intimate of *Brahms*.

d.1951

James Bridie, English dramatist and physician, author of *The Anatomist*. His *Dr. Angelus* was based on the career of the physician, *Dr. Edward William Pritchard*, who murdered his wife, mother-in-law, and mistress.

January 30

1577

Vicary's Anatomy of Man is published.

d.1784

Thomas Falkner, English physician, botanist, and pharmacist. He started his career as a surgeon on a slave ship and ended his days as a Jesuit missionary.

Vicary

d.1888

Asa Gray, American botanist and physician. *Lilium Grayii* is but one eponym commemorating this Darwinian biologist; his name is remembered in three genera, many species, and a mountain peak in California. The last was named by another physician botanist, *Charles Christopher Parry* (1823–1890), who in turn is remembered in *Lilium Parryi* and a mountain peak in Colorado.

January 31

1901

The Russian physician playwrite *Anton Chekhov's Three Sisters* opens to mixed reviews in Moscow; Ms. Chekhov plays Masha.

1953

Pravda announces the discovery of the Doctor's Plot, an alleged conspiracy by Russian physicians to murder Soviet military leaders.

d.1836

John Cheyne, Scottish physician. He studied diseases of children and gave the first description of acute hydrocephalus (1808). With the Irish physician, *William Stokes,* he characterized a variety of disorders of respiration.

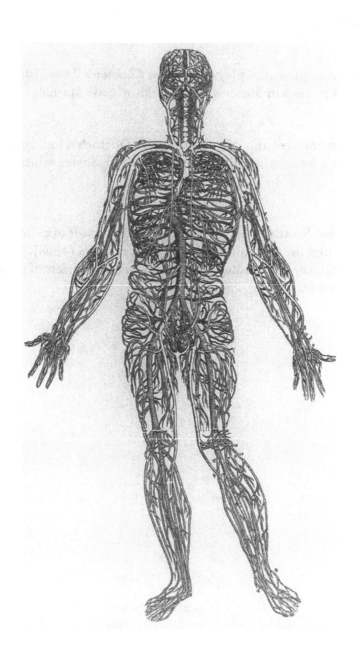

February

> *While slant sun of February pour*
> *Into the bowers a flood of light*
> > —*William Cullen Bryant*, *"A Winter Piece"*

February 1

Feast of *Saint Bridget* (453–525) of Ireland; she practiced medicine and midwifery and persuaded rulers to banish quacks.

b.1716

John Bard, American physician and surgeon. In 1759 he performed the first successful operation for an ectopic pregnancy. He was the first president of the Medical Society of the State of New York.

b.1794

Johann Friedrich Dieffenbach, German surgeon and botanist. In 1820 he reintroduced rhinoplasty and in 1839 he performed the first successful strabotomy (to correct strabismus). He is remembered in the name of dumb cane, *Dieffenbachia picta.*

b.1801

Emile Littré, French physician and philologist. He compiled a Greek dictionary, edited *Pliny,* and published the landmark 10-volume *Oeuvres d'Hippocrate* (1839–1861).

1941

First attempt to use penicillin to treat an advanced staphylococcal infection; the patient died when the limited supply of the experimental drug ran out.

February 2

1677/8

Thomas Thacher, a doctor–preacher of Boston, publishes a one-page pamphlet that is the first medical book in America: *A Brief Rule to Guide the Common People of New-England How to Order Themselves and Theirs in the Small Pocks or Measles* (Boston: John Foster).

1710

Dr. Thomas Dover rescues *Alexander Selkirk,* who had been marooned on Juan Fernandez island. *Selkirk* is the original Robinson Crusoe— "I am monarch of all I survey. My right there is none to dispute" *(Cowper).* This English buccaneer was also noted for his *Dover's powders* or *pulvis Ipecacuanhae compositus,* a mixture of ipecac, opium, and milk sugar. He was also known as "Doctor Quicksilver" because he advocated metallic mercury for syphilis and a variety of other disorders. Despite his pharmaceutical contributions, he was not addicted to prescribing—"I never affronted an apothecary, unless ordering too little physic, and curing a patient too soon, is in their way of thinking, an unpardonable crime." "A good fighter, a good hater, as alas! so many physicians have been" *(Osler).*

d.1723

Antonio Maria Valsalva, Italian anatomist, one of the founders of otology. His *De aure humana tractatus* (Bologna, 1704) first divided the ear into three parts—inner, middle, and outer. He succeeded *Malpighi* at Bologna, and *Morgagni* was one of his students. He described the aorta, the colon, the vagus nerve, and the eponymic maneuver in which intrapulmonic pressure is increased by forcible exhalation against a closed glottis.

1884

Robert Koch discovers the cholera vibrio. This country doctor—"our medical Galileo" *(Osler)*—was awarded the 1905 Nobel prize for his work in bacteriology.

February 3

Feast of **Saint Blaise**, Bishop of Sebaste in Armenia, patron saint of those suffering diseases of the throat. *St. Blaise's disease* is quinsy, sore throat, or tonsillitis. He is also the patron saint of woolcombers and wild animals. An incident when he saved a child from choking on a swallowed fishbone suggests his patronage of the Heimlich maneuver.

b.1777

John Cheyne, Scottish physician.

b.1874

Gertrude Stein, American writer. She was a medical student at Johns Hopkins from 1897 to 1902. "Now listen! I'm no fool. I know that in daily life we don't go around saying 'is a...is...is...' Yes, I'm no fool; but I think that in that line the rose is red for the first time in English poetry for a hundred years." *Hemingway* quipped, "Gertrude Stein and me are just like brothers."

b.1892

Sir Morell Mackenzie, one of the founders of modern otolaryngology. He built Golden Square Hospital, the first hospital exclusively devoted to diseases of the throat. This London operation was very high volume: "If I am ever to make anything of the throat, I must see more patients." In a letter of recommendation, *Queen Victoria* wrote, "*Sir William Jenner* said *Dr. M. Mackenzie* certainly is very clever in that particular line of throats; but that he is greedy and grasping about money and tries to make a profit out of his attendance." Unfortunately, this was in reference to the infamous missed diagnosis of cancer in the *Emperor Friedrich III.*

February

February 4

d.1884

George Engelmann, American physician, botanist, and pioneer meteorologist. He founded the Saint Louis Academy of Science and is remembered in the eponyms for three plant genera, a large number of botanical species.

b.1902

Charles Lindbergh, American aviator and inventor, the "Lone Eagle." In 1936 he collaborated with *Alexis Carrel* to design the *Lindbergh perfusion pump*, an early artificial heart mechanism that allowed organs to be kept alive outside the body.

Dr. and Mme. Carrel with Charles Lindbergh, 1936

b.1905

Carl Muschenheim, discoverer of the isoniazid, or INH, treatment for tuberculosis.

February 5

Feast of *Saint Agatha*, patron all those suffering of diseases of the female breast as well as patron saint of nursing women.

b.1745

John Jeffries, author of *A Narrative of Two Aerial Voyages*, 1786.

b.1866

Sir Arthur Keith, British anatomist and anthropologist. He identified the natural pacemaker of the heart, the *sinoatrial node of Keith and Flack.* He was one of the more prominent scientists who were taken in by the Piltdown Skull hoax.

February 6

b.1892

William Murphy, American physician and medical researcher. He was awarded the 1934 Nobel Prize for his work in developing the use of liver in the treatment of pernicious anemia.

February 7

1837

Florence Nightingale: "For what training is there compared to that of the Catholic nun?" God called her to His service February 7, 1837.

1901

Paul Uhlenhuth publishes a paper describing a precipitin test to identify human blood stains.

Florence Nightingale

February 8

b.1577

Robert Burton, English author of *The Anatomy of Melancholy* (1621), one of the first psychiatric encyclopedias.

> *"All my joys to this are folly*
> *Naught so sweet as Melancholy"*

Anthony à Wood said of him, "he ... paid his last debt to Nature, in his Chamber at Christ Church at, or very near that time, which he had some years before foretold from the calculation of his own nativity, which being exact, several of the students did not forebear to whisper among themselves, that rather than there should be a mistake in the calculation, he sent up his soul to heaven through a slip about his neck."

February 9

Feast of *Saint Apollonia*, martyr, aged deaconess of Alexandria. She is the patron saint of those afflicted with dental diseases and toothache.

b.1798

William Draper Brincke, American physician and pomologist. He was a founder of the *American Pomological Society*.

Saint Apollonia

February 10

Feast of *Saint Scholastica*, virgin, sister of *Saint Benedict*. She helped the sick during the plague in Italy and established hospitals and trained nurses "to bathe the sick, give them medicine and food, and pray with the dying."

d.1878

Claude Bernard, French physiologist.

d.1912

Joseph Lister, an English surgeon. "To intrude an unskilled hand into such a piece of Divine mechanism as the human body is indeed a fearful responsibility."

b.1920

Alex Comfort, poet, gerontologist, and sexologist.

d.1923

Wilhelm Konrad Roentgen

Wilhelm Konrad Roentgen, German radiologist.

February 11

Feast of *Saint Bernadette Soubirous*, asthmatic; she received an appar-
ition of the Virgin at Lourdes which has since become one of the
principal healing shrines in modern Christendom. She served as a
nurse in the Franco-Prussian War.

1144

Robert of Chester finished his Latin translation of an Arabic work on
alchemy, the first of its kind to reach Latin civilization.

1752

First patients are admitted to the earliest American hospital, the
Pennsylvania Hospital, Philadelphia, founded to serve the "distem-
pered Poor."

d.1937

Adolf Gaston Eugen Fick, Swiss German ophthalmologist who devised
the first practical contact lens.

February 12

b.1567

Thomas Campion, Elizabethan lyric poet and composer, and a Catholic convert with degrees in law and medicine. He was an outstanding song writer of "ayres" and not madrigals.

1828

Dr. James Blundell performs the first successful hysterectomy. He was
a founder of modern abdominal surgery and pioneered in the use
of blood transfusion by the indirect syringe method.

February 13

1832

The first appearance of Asiatic cholera in London at Limehouse
and Rotherhithe was noted.

1843

Oliver Wendell Holmes reads his paper "On the Contagiousness of Puerperal Fever" to the Boston Society for Medical Improvement.

1922

The Royal Society of London receives "On a Remarkable Bacteriolytic Element Found in Tissues and Secretions" by *Alexander Fleming*.

1929

A pioneering paper, "Culture of a Penicillium," is read to the Medical Research Club by *Alexander Fleming*.

February 14

Feast of *Saint Valentine*. *Saint Valentine's disease* is epilepsy, which is also *St. Avertin's disease, St. John's evil,* and *St. Mathurin's disease.*

b.1728

John Hunter, Scottish physician, one of the greatest surgeons of all time, dentist, collector, and master physiologist. "He was, in the first place, a naturalist to whom pathological processes were only a small part of a stupendous whole, governed by law, which, however, could never be understood until the facts had been accumulated, tabulated and systematized....He made all physicians naturalists." *Osler* said that he combined the "qualities of *Vesalius, Harvey,* and *Morgagni* in an extraordinary personality." He studied the function of the air sac in birds, the olfactory nerves, sexual behavior in animals, comparative aspects of the placenta and the testes, the behavior of bees, and temperature variations in plants and animals. "He was almost adored by the rising generation of medical men who seemed to quote him as the Schools, at one time, did *Aristotle.*"

1852

Eliza Armstrong, age three and a half, is the first pediatric patient admitted to the first children's hospital, Great Ormund Street.

b.1911

Willem Johan Kolff, Dutch physician. He built the first artificial kidney dialysis machine; it was first used in 1944 on Dutch resistance fighters. The first patient whose life was saved was Sophia Schafstadt, a Nazi sympathizer.

February 15

b.1755

Jean Nicolas Corvisart des Marets, physician to *Napoleon.* He revived *Auenbrugger's* lost art of percussion. "Medicine is a conjectural art." He wrote *Essai sur les maladies et les lésions organiques du coeur et de gros vaisseaux* (1806). *Corvisart's disease* is chronic hypertrophic myocarditis, and *Corvisart's facies* is that of cardiac insufficiency.

b.1829

Silas Weir Mitchell, American Civil War surgeon, neurologist, and popular novelist. The founder and spiritual leader of American neurology, he described posthemiplegic chorea in 1874, the relationship between pain and weather in 1877, and the relationship of eye strain (astigmatism) to headache. *Mitchell's* treatment of bed rest, isolation, and a nourishing diet to treat neurosis, was put forth in *On Rest in the Treatment of Nervous Disease* (1875). His novels exhibited pioneer observations in psychiatry,

Silas Weir Mitchell

but he complained, "Ever since the Crimean War, nurses have been getting into novels." *Osler* said of him, "If asked for a scroll to place beneath [his portrait], I would write that he was one

> *Whose even balanced soul*
> *Business could not make dull, nor passion wild;*
> *Who saw life steadily and saw it whole!"*

b.1853

Frederick Treves, a pioneering English abdominal surgeon. He wrote traveler's tales, discovered the "Elephant Man," and wrote a volume of *Browning* criticism, *The Ring and the Book* (1913). *Treves' bloodless fold* is the eponymic plica ileocecalis of the appendix.

February 16

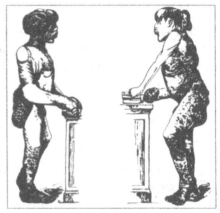

Two views of the "Elephant Man"

b.1822

Francis Galton, English explorer, anthropologist, and eugenicist. He was a founder of the statistical school of genetics and the study of heredity. He wrote *Hereditary Genius* (1869), formulated the law of filial regression to the mean as well as *Galton's* law in which an ancestor of the nth degree contributes $2\,(1/2)^n$ of the heritage. He pioneered in the use of twins in genetic research and in the new science of dactyloscopy or the study of fingerprints. He devised the ALW (arch-loop-whorl) classification system and described *Galton's delta,* a triangular area of papillary ridges on the distal pad of the digits. He also founded modern weather mapping techniques. Cape Hyacinth, *Galtonia candicans*, is named for him.

d.1857

Henry David Ficinus, German physician, physicist, pharmacist, botanist, and optician.

d.1857

Elisha Kent Kane, American surgeon, Arctic explorer. Known as "Kane of the Arctic," his *Arctic Explorations* (1856) was one of the most widely read books of the nineteenth century. *Kane Basin* in the Arctic Ocean is named for him.

Claude Bernard

d.1878

Claude Bernard, French physiologist. He was discharged at the age of twenty from his position in a pharmacy because he was a daydreamer. When he finally gave up his hopes of a career as a dramatist, he developed the concepts of homeostasis and the milieu intérieur, discovered the glycogenic function of the liver, the role of pancreatic enzymes in digestion, vasomotor control of blood flow, and various aspects of carbon monoxide poisoning. This aloof genius was the first scientist honored by a funeral at public expense.

February 17

b.1688

Cadwalader Colder, American physician, philosopher, historian, naturalist, and Lieutenant Governor of New York (1761). *Coldenia procumbens* is named for him.

1735

William Douglas (1691–1752) writes to *Dr. Colder:* "We have lately in Boston formed a medical society."

b.1781

René Theophile Hyacinthe Laënnec, French physician. His *De l'Auscultation Médiate* (1819) records his discovery of the stethoscope. *Laënnec's cirrhosis* is portal cirrhosis. *Thomas Addison* said of him, "*Laënnec* contributed more toward the advancement of the medical art than any other single individual."

b.1845

Charles McBurney, New York surgeon. This pioneer abdominal surgeon identified the point of maximal tenderness in appendicitis—*McBurney's point.*

February 18

b.1896

André Breton, French physician, poet, and essayist. He had a special interest in mental disease and drew on Freudian dogma to become the founder and leading theorist of Surrealism with its reliance on dreams and "automatic writing." His poem "Noeud des miroirs" includes a prescription formula for angina pectoris.

February 19

b.1473

Nicolaus Copernicus, Polish astronomer. He studied mathematics, astronomy, and law at Cracow, Bologna, and Ferrara; he took his medical degree at Padua in 1501.

b.1660

Friedrich Hoffmann, German physician. His family was connected with medicine for 200 years before him. In 1693 he founded the University of Halle and took the chair of medicine there. He viewed pathology as an aspect of physiology; as a mechanist he saw therapy as sedative, corroborant, tonic, or evacuant. He was an early and enthusiastic advocate of crenology—the therapeutic use of mineral waters.

b.1834

Hermann Snellen, Dutch ophthalmologist. In 1862 he devised a test for central visual acuity that bears his name (the *Snellen Chart*).

1878

Louis Pasteur argued the germ theory of infection before the French Academy of Medicine.

February 20

b.1816

William Rimmer, American shoemaker, physician, painter, and sculptor. His *Art Anatomy* is still used today.

d.1844

Samuel Fowler, American physician. He discovered two rare metal ores named Fowlerite and Franklinite, and named them for himself and the New Jersey town in which he practiced.

February 21

1866

Lucy Hobbs Taylor receives her DDS to become the first woman dentist.

d.1941

Frederick Grant Banting, Canadian orthopedic surgeon, codiscoverer with *Charles H. Best* of insulin. A pioneer of aviation medicine, he was killed in the crash of a Canadian bomber off Newfoundland. He won the Nobel Prize in 1923.

Best and Banting

February 22

d.1794

Kaspar Friedrich Wolff, a Russian physician and anatomist who was one of the founders of modern embryology. He was responsible for the doctrine of the germ layers and is eponymously remembered in a number of renal primordia.

d.1818

Archibald Bruce, American physician and mineralogist. A magnesium hydroxide he discovered in New Jersey—brucite—is named for him. "His appearance being that of one who enjoyed his food and drink, he died of apoplexy at 41."

1828

The German chemist *Friedrich Wöhler* (1800–1882) announces, in a letter to *J. J. Berzelius,* the isolation of urea, thus demonstrating that an organic substance could be synthesized from non-organic components. This marked the beginning of the study of metabolic pathways.

February 23

1672

Official writ appointing *Niels Stensen* Royal Anatomist of Denmark.

b.1801

James Hope, English surgeon. His popular *Treatise on Diseases of the Heart and Great Vessels* (1831) contained many original observations on valvular stenosis and aortic regurgitation, and anticipated the description of *Corrigan's pulse.*

d.1821

John Keats, "Here lies one whose name was writ in water."

b.1823

John Braxton Hicks, English gynecologist. He introduced podalic version (internal and external) and described the contractions that bear his name—irregular painless contractions of the uterus after the first trimester of pregnancy, persisting to term.

1946

Edmund Lohr, Jr., of Kew Gardens, Queens, becomes the first known victim of rickettsial pox.

1954

The first mass polio immunization of school children takes place in Pittsburgh.

February 24

1461

Charter of corporation for Masters or Governors of the Mystery of London (barber–surgeons).

d.1636

Santorio Santorio (*Sanctorius*), Italian physician. He used a thermometer (balance thermoscope) to measure body temperature and counted the pulse with a pulsimeter.

b.1876

Herbert Spencer Dickey, an American physician and South American explorer.

1960

Librium approved by the FDA.

February 25

b.1628

John Baptist Morgagni, Italian physician, pathologist, and anatomist. His *De sedibus et causis morborum per anatomen indagatis* was published in Venice in five volumes in his eightieth year. He showed that diseases arise from definite changes in some organ of the body and these changes are constant for any particular disease. He was the father of the "Anatomical Concept" of disease. "And into this metaphysical confusion *Morgagni* came like an old Greek with his clear observation, sensible thinking, and ripe scholarship." *(Osler)* He was a classical scholar, a medical biographer (of *Valsalva*), and is remembered in numerous eponyms.

b.1908

Frank Gill Slaughter, an American surgeon and novelist, author of *Women in White* and *Sword and Scalpel.*

February 26

d.1693

Charles Scarburgh, English physician and mathematician. He worked with *Harvey* in his experiments on the generation of animals.

b.1849

Franz Christian Boll, anatomist, professor of physiology at Rome. His *Sull'anatomia e fisiologia della retina* (1876) described the visual purple (rhodopsin) in the rods of the retina that fades in the presence of light. *Boll's cells* occur in the lacrimal and mammary glands.

d.1884

Alexander Wood, Scottish physician. He introduced the use of a hypodermic syringe for the administration of drugs.

d.1903

Richard Gatling, American physician and inventor. He is best remembered for the machine gun that bears his name.

d.1917

Joseph Jules Déjérine, French neurologist. He married one of his American medical students, the beautiful *Augusta Klumpke.* Their names are frequently together in the eponyms for many neurological conditions. *Klumpke's paralysis* is palsy and atrophy of the hand muscles secondary to a brachial plexus lesion suffered at birth.

d.1969

Karl Jaspers, psychiatrist and existential philosopher. "Without God, only idols remain."

February 27

d.1735

John Arbuthnot, British physician, mathematician, wit and man of letters. He was a friend to *Swift* and *Pope,* one of the leading spirits of the Scriblerus Club and author of the *Memoirs of Martin Scriblerus.* His *History of John Bull* created that character. He was royal physician to **Queen Anne,** "a man who during a long life...while classic-serving, antiquarianising, science-seeking, satire-writing, wit-making, and fun-distributing, managed, by hook or by crook, to write prescriptions, the physic of which the people and even royalty swallowed...with infinitely more faith in its efficacy than ever satisfied the conscience of the renowned prescriber." *(Richardson)*

b.1899

Charles Best, codiscoverer of insulin, winner of 1923 Nobel Prize.

d.1936

Ivan Pavlov, Russian physician and psychologist, winner of the 1904 Nobel Prize in Physiology for his work on reflexes.

February 28

1873

Gerhard Hansen discovers *Mycobacterium leprae,* the cause of leprosy.

1896

Bernard Jean Antonin Marfan (1858–1942), Parisian physician, publishes the first report of a syndrome that will bear his name, *Marfan's disease*—arachnodactyly with bilateral ectopia lentis.

b.1896

Philip Showalter Hench, American physician who pioneered in the use of cortisone for arthritis. He had noticed that women with arthritis had less pain when they were jaundiced, pregnant, or undergoing surgery. He shared the Nobel Prize with **E.C. Kendall** and **T. Reichstein** in 1950.

b.1901

Linus Pauling, American chemist who pioneered in the field of molecular disease. He applied his often controversial new concepts to sickle cell disease, megavitamin therapy, and the use of vitamin C to prevent the common cold. He won the 1954 Nobel Prize for his work in chemistry and the 1962 Nobel Peace Prize.

d.1769

Antoine Ferrein, French surgeon and anatomist. He originated the term "vocal cord" and compared the ligaments of the larynx to violin strings and the air passing through to the bow.

d.1929

Clemens von Pirquet, the Austrian pathologist who coined the word "allergy."

February 29

b.1820

Lewis Albert Sayre, "Father of American Orthopedic Surgery." He was one of the founders of Bellevue Hospital Medical College in 1859 and is remembered for his development of *Sayre's jacket* of plaster of Paris to treat scoliosis and *Pott's disease.*

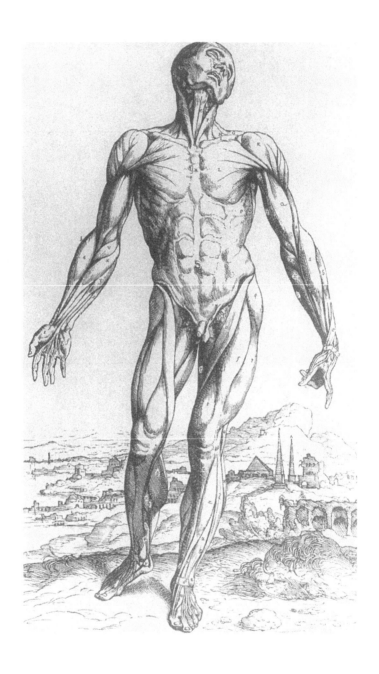

March

March 1

1648

"*Henry Ivatt* complained against *Anthony Mold* for his evil practice on the wife of the said Ivatt who being afflicted with the King's Evil whereof he undertook to cure her. And for that purpose did receive of the said Ivatt thirty shillings in hand and was to have forty shillings more when she was cured; both parties referred themselves to this Court, whereupon this Court doth order that the said *Mold* do restore twenty shillings back again to the said *Ivatt*, which he promised to pay accordingly and so all differences between the said parties by their own consent to cease and determine." (*Records of the Barber Surgeons of London*)

d.1667

Francesco Redi of Arezzo, Italian physician, naturalist, and poet, the "Father of Helminthology." He disproved the spontaneous generation of maggots from decayed meat and studied viper poisons. He was one of the last of the Tuscan dialect poets; the meter of his *Bacco in Toscana* mimics the increasing intoxication of the wine-tasting Bacchus.

1907

Alois Alzheimer (1864–1915), a German neurologist, described the presenile degeneration that is now known by his name.

March 2

d.1840

Heinrich Wilhelm Matthäus Olbers, German physician and astronomer. He studied asteroids and devised Olbers' method of calculating the orbits of comets. He caused considerable scientific controversy by asking the basic question, Why is the night sky dark? (*Olbers' paradox*).

b.1841

Adolph Pansch, German anatomist. In 1869 he went on an expedition to the North Pole. The sulcus intraparietalis is named for him.

March 3

d.1616

Matthias de l'Obel (b. 1538) Flemish physician and botanist who collected data on 1200 plants and grouped families by leaves. His rough notions of genus and species were later developed by *Linnaeus* who named the family Lobeliaceae and the genus *Lobelia* after him.

1781

James Craik is appointed chief physician and surgeon of the Continental Army. He attended *George Washington* in his final illness.

1816

John Keats registers as dresser to surgeon–teacher *Billy Lucas*, one of the most ignorant practitioners in the history of the art.

1824

French surgeon *Jean Civiale* completes the first lithotripsy (breakup of a bladder stone *in situ*).

March 4

b.1844

Thomas Lauder Brunton, English physician. A founder of modern pharmacology, he was the first to use amyl nitrite for angina (1867). He also invented an otoscope.

b.1855

Luther Emmett Holt, American pediatrician. He authored the classic textbook, *The Diseases of Infancy and Childhood* (1897) and introduced the term "heparin."

d.1952

Charles Scott Sherrington, English neurophysiologist. His classic work is *The Integrative Action of the Nervous System* (1906). He received the 1932 Nobel Prize in Physiology.

d.1963

William Carlos Williams, American pediatrician and poet. His works include *Journey to Love* and *Paterson.* On himself: "I don't play golf, am not a joiner. I vote Democratic, read as much as my eyes will stand, and work at my trade day in and day out. When I can find nothing better to do, I write."

March 5

d.1763

William Smellie, English midwife and village practitioner. He measured for contracted pelvis, invented a scissors for craniotomy, and wrote *A Treatise on the Theory and Practice of Midwifery* (1752). "A great horse-godmother of a he-midwife."

d.1815

Franz Mesmer, German physician and hypnotist. His "magnetic therapy" included an imposition of hands on the patient.

March 6

1664

First publication of the *Philosophical Transactions.*

b.1740

Giovanni Meli, Sicilian physician and poet. He was professor of chemistry at Palermo. As the "Sicilian Anacreon," he gave prestige to the Sicilian dialect. His *Don Chisciotti e Sanciu Panza* was a satirical continuation of *Cervantes'* novel.

b.1885

William Norman Pickles, an English "country doctor" who studied infectious diseases in small communities. He wrote *Epidemiology in Country Practice* (1939).

d.1888

Louisa May Alcott, American author. Her *Hospital Sketches* (1863) is composed of letters to her family while she was a nurse in a Union hospital in Georgetown during the American Civil War.

1900

The first United States bubonic plague epidemic begins in San Francisco.

d.1971

Herbert McLean Evans, American investigator. He discovered vitamin E and isolated pituitary growth hormone.

March 7

b.1792

Sir Thomas Watson, English physician. In 1843 he suggested the use of rubber gloves in surgery.

d.1863

Charles W. Short, Kentucky physician and botanist. Oconee-bells are named after him—*Shortia galacifolia.*

d.1969

Sadao Otoni, pathologist. He identified an eosinophilic granuloma of bone and described the glomus jugulare tumor of the middle ear (*Otoni's tumor*).

March 8

Feast of *Saint John of God* (1495–1550), Portuguese shepherd, bookseller, and founder of the Brothers Hospitallers. With *Saint Camillus de Lellis* he is patron saint of nurses, hospitals, and the sick. He is also the patron saint of booksellers and printers.

b.1666; d.1709

William Cowper, English surgeon and anatomist. He wrote a classic description of aortic insufficiency and the glands of the membranous urethra are named for him. His *Anatomy of Human Bodies* (1698) contained mostly plagiarized drawings from *Govert Bidloo.*

William Cowper

b.1752

Johann David Schopf, German surgeon. While serving in the British Army during the American War of Independence, he collected material that later became seminal treatises on American geology, climate, diseases, ichthyology, frogs, and turtles.

b.1884

Anthony Lanza, was an American physician who discovered silicosis.

March 9

1610

Thomas Lodge is admitted to the Royal College of Physicians. *Lodge* (1557–1625) was an Elizabethan poet, lawyer, and physician. He started out as a privateer in the New World and became possibly the earliest English satirist. His *Rosalynde* (1590) gave *Shakespeare* the story for *As You Like It,* and the idea for *Venus and Adonis* may have arisen from Lodge's *Scillaes Metamorphosis* or *Glaucus and Scilla.* His translations are some of the best in the language.

b.1826

Georges Fernand Isidore Widal, French physician. He developed a serodiagnostic test for typhoid fever and described a type of the disease called *Hayem-Widal chronic hemolytic jaundice.*

1907

The Indiana legislature passes the first eugenic sterilization law in the United States.

March 10

b.1628

Marcello Malpighi, Italian anatomist and microscopist. He made many contributions to embryology, botany, and entomology, discovered erythrocytes and the capillary circulation. He is remembered in many anatomical eponyms. Opposition by the Galenists was so heated that in 1689 he was assaulted by two masked university colleagues.

Marcello Malpighi

1980

Millionaire *Scarsdale Diet* doctor **Herman Tarnower** is shot to death by his former mistress, the headmistress of an exclusive girls school, *Jean Harris*. She was found guilty of second degree murder on February 24, 1981.

March 11

1818

Publication date of **Mary Shelley's** *Frankenstein,* a morality tale anticipating modern transplant medicine.

d.1913

John Shaw Billings, American surgeon and librarian. He designed the Johns Hopkins Hospital and with **R. Fletcher** founded the *Index Medicus.* "The education of the doctor which goes on after he has his degree is, after all, the most important part of his education." "A hospital is a living organism."

d.1955

Alexander Fleming, a Scottish bacteriologist and the discoverer of penicillin.

March 12

1799

Dr. Benjamin Waterhouse writes "Something Curious in the Medical Line," the first account of vaccination in America.

1845

Francis Rynd first introduces fluids into the body by subcutaneous injections using a hypodermic syringe.

d.1853

Mathieu-Joseph-Bonaventure Orfila, Spanish physician and chemist. He was the father of medicolegal toxicology.

b.1929

William Liley, "Father of Fetology." He adopted a Down syndrome child who was otherwise to be institutionalized.

March 13

d.1824

J.A.M. Méglin, French physician and anatomist. He first described *Méglin's palatine point* of the emergence of the descending palatine nerve from the palato-maxillary canal.

d.1888

Claude P. Blot, French obstetrician. He invented *Blot's scissors* to perform a craniotomy in difficult deliveries. This is distinct from *Pajot's hook* which was used to decapitate the fetus by the sawing movement of a string in the hook's groove. The latter instrument was devised by Blot's contemporary, *Charles Pajot.*

Accouchement by a bull

March 14

b.1845

Paul Ehrlich, German bacteriologist. He was the founder of the principles of modern chemotherapy and a pioneer in the staining techniques for cells and bacteria. He received the 1908 Nobel Prize for his work.

March 15

44 BC

The Roman physician *Antistus* examines the body of *Julius Caesar* and concludes that of the 23 stab wounds only the wound in the chest was mortal.

b.1881

DeForest Clinton Jarvis, American physician and author of *Folk Medicine: A Vermont Doctor's Guide to Good Health* (1958).

1937

First blood bank opened.

March 16

d.1825

Pierre Augustin Béclard, Parisian anatomist who described *Béclard's triangle*—the area bounded by the posterior border of the hypoglossus, the posterior belly of the digastric, and the greater cornu of the hyoid bone.

1867

Joseph Lister's original paper on carbolic acid antisepsis is published in the *Lancet:* "On a New Method of Treating Compound Fractures, Abscess, &c., with Observations on the Conditions of Suppuration."

d.1935

John James Macleod, Scottish physician. With **Banting** and **Best** he discovered the insulin treatment for diabetes.

March 17

b.1741

William Withering, English physician, scientist, translator, and mineralogist. He wrote the classic *An Account of the Foxglove* (1785) and discovered the mineral now named in his honor, Witherite.

1843

At a St. Patrick's Day party, **Charles Dickens** makes the acquaintance of *Dr. Miles Marley,* a practitioner of Cork Street, Picadilly. With regard to the to the latter's unusual surname, **Dickens** comments that "your name will be a household word before the year is out."

d.1957

Henry Ernest Sigerist, physician and medical historian. "Medicine is basically a social science."—*Civilization and Disease.*

March 18

b.1721

Tobias Smollett, Scottish physician, surgeon, translator, playwright, novelist, historian, and editor. His novels (*Roderick Random, Peregrine Pickle,* and *The Expedition of Humphrey Clinker*) and his translations (*Gil Blas,* and *Don Quixote*) are still read.

b.1818

Joseph von Lenhossek, Hungarian anatomist. He, his father, and his son wrote much on medical subjects, especially the central nervous system, and they are remembered by many eponymous structures.

b.1864

Carl Schlatter, Swiss surgeon. He performed the first gastrectomy in 1897 and in 1903 described the painful lesions of the tibial tuberosity that became known as *Osgood-Schlatter's disease.*

March 19

1540

"In Ferrara Italy was born a monster as big and well formed as if it were four months old, having both feminine and masculine sexual organs and two heads, the one of a male and the other of a female." **Caelius Rhodiginus**

The 'monster' of Ferraro

b.1774

Franz Paula von Gruithuisen, German physician and astronomer. He worked on lithotripsy, an operation to crush stones in the bladder or urethra, and proposed that the craters on the moon's surface were caused by meteors.

d.1816

Philip Mazzei, Italian physician, surgeon, merchant, and horticulturist. In 1773 he introduced the cultivation of grapes and olives to Virginia. He exerted a strong influence on the Italian tastes of his close friend, *Thomas Jefferson.*

b.1821

Richard Francis Burton. This Victorian polymath (explorer, translator, linguist, archeologist, anthropologist, poet, soldier, diplomat, and sexologist) preferred to travel in disguise. In his assumed identity of the Afghan physician *Pathan* he practiced medicine (cathartics and hypnosis) on the slave girls of Cairo and in other cities of the east.

March 20

1346

Venice forms a Council of Three to segregate shipping suspected of harboring infection. The physician and historian, *Simon de Covina,* is a prime mover in setting up what is first a thirty day (*trentina*) isolation, but soon is changed to a forty day period (or *quarantino*) because this was the period that both *Moses* and *Christ* spent in the desert.

b.1728

Simon Auguste André David Tissot, Swiss physician. He wrote on nervous diseases and hygiene and recommended exercise for children. His most famous work was *L'Onanisme* (1760).

William Richard Gowers,

b.1845

William Richard Gowers, Victorian physician and polymath, one of the founders of modern neurology. He invented *Gowers' hemoglobinometer, Gowers' hemocytometer, Gowers' safety syringe, Gowers' magnifying otoscope,* a portable reading lamp, a home intercom system, a three-wicked candle, and was a passionate enthusiast of the use of shorthand. His name is eponymously attached to the anterior spinocerebellar tract *(Gowers' tract),* irregular pupillary light reflex as a sign of tabes dorsalis, the sign of the patient climbing up on himself in muscular dystrophy or other disorders of large proximal muscle involvement, and *Gowers' disease of saltatory spasms.* It was his authority that gave preference to the term "knee jerk" over "patellar reflex." His *Textbook of Neurology* remains the "Bible" in the field.

d.1853

Robert James Graves, Irish physician. He was a leader of the Dublin school of medicine in its prime. *Graves' disease* is exophthalmic goiter. He suggested as his epitaph, "He fed fevers."

d.1878

Julius Robert Mayer, German physician, physiologist, and physicist. As a ship surgeon on an 1840 voyage to Java, he made physiological observations that led him to formulate the Law of the Conservation of Energy in 1842.

Robert James Graves

March 21

Feast of *Saint Benedict*, patron saint of all suffering disease.

1850

Doctor *Oliver Wendell Holmes*, Parkman Professor of Anatomy at Harvard Medical School, testifies on the possible identity of the professionally dismembered remains of *Dr. George Parkman* (his professorship was named after the victim) who was murdered by *Dr. John Webster*, Professor of Chemistry and Mineralogy at Harvard, in an argument over money.

1935

The first run of an incubator ambulance service for prematures in Chicago.

March 22

1800

The *Royal College of Surgeons* is established. The *Lancet* would later describe this body as "this sink of infamy and corruption, this receptacle of all that is avaricious, base, worthless, and detestable in the surgical profession."

d.1913

Joseph Georg Egger, German physician, geologist, and micropaleontologist with a special interest in foraminifers and diatoms.

March 23

b.1501

Pietro Andrea Mattioli, Italian physician and botanist. He translated *Dioscorides* in 1544 and his *Commentary on Dioscorides* (1554) is an encyclopedia of Renaissance pharmacology. He was a venerated physician who stayed to fight epidemics and died of the plague. He is remembered in the name of the Brompton Stock or gilliflower—*Matthiola incana*.

1930

The Russian surgeon *Sergei Yudin* performs the first transfusion of cadaver blood into a human.

March 24

1345

The conjunction of Saturn, Jupiter, and Mars in the 40th degree of Aquarius is recorded as the cause of the Black Death of 1348 according to *Guy de Chauliac's Cirurgia* (1363). *De Chaulaic* was the "Father of Surgery" according to *Fallopius,* and his textbook was the fullest medieval survey on the subject of surgery. He remained on duty during the plague: "So great was the contagion that one caught it from the other not only by being with him, but by simply looking at him."

d.1571

Bartolomeo Maranta, Venetian physician and botanist. He is remembered in the name of the prayer plant—*Maranta leuconeura.*

d.1875

Daniel Hanbury, English pharmacist. The Hanbury Memorial Medal is awarded for original research in pharmacy.

1882

In his paper, "The Etiology of Tuberculosis," *Robert Koch* announces to the Berlin Physiological Society the discovery of the *Myobacterium tuberculosis.*

March 25

1525

The first English herbal is published anonymously by *Richard Banckes.*

1763

New York physician *James Jay* is knighted. Later, during the American War of Independence, he invents an invisible ink to be used in secret correspondence. When he is accused of treason, his brother *John Jay* comments, "If after making so much bustle in and for America, he has, as it is surmised, improperly made his peace with Britain, I shall endeavour to forget that my father has such a son."

d.1712

Nehemiah Grew, English physician, botanist, and microscopist. With *Malpighi* he is considered the founder of plant anatomy with his *Anatomie of Plantes* (1682). His *Cosmologia sacra* (1701) is a landmark teleological treatise. He also compiled a catalog of rarities in the Museum of the Royal Society. He coined the term "comparative anatomy." *Linnaeus* named the genus of trees *Grewis* in his honor.

d.1889

Claude-Adolphe Nativelle, French pharmacist who was the first to isolate digitoxin.

1985

Dr. Haing S. Ngor wins an Oscar as best supporting actor in *The Killing Fields.*

March 26

1669

Dated English manuscript of *Thomas Sydenham's* classic volume, *Medical Observations.*

b.1516

Konrad von Gesner, Swiss physician, botanist, zoologist, linguist, and bibliographer. With his *A Catalogue of Animals* he became known as the "German Pliny"; his 20 volume *Bibliotheca Universalis* (1545–1549) was the earliest biographical dictionary and earned its author the title of "Father of Bibliography." He was one of the first real mountaineers; as a young man he vowed to climb at least one mountain every year.

d.1861

Andrew Sinclair, British naval surgeon and naturalist. The name *Sinclaira* was given to a tropical American genus of Compositae (now merged with Liabun).

March 27

b.1845

Wilhelm Konrad Röntgen, German radiologist. "The question 'What would your mother have said and done in this or that confused situation?' has often shown me the right way out of it."

1914

A. Hustin of Belgium performs the first successful indirect human blood transfusion at l'Hôpital Saint-Jean in Brussels. He had previously discovered that citrate used to prevent clotting was nontoxic and did not significantly lower the calcium level.

March 28

b.1799

Karl von Basedow, German physician. *Von Basedow's disease* is exophthalmic goiter.

d.1936

Sir Archibald Edward Garrod, English physician and pioneer in the area of inborn errors of metabolism. He studied the genetics of alkaptonuria, fructosuria, and albinism.

March 29

1886

Dr. John Pemberton, a chemist from Atlanta, Georgia, launches an "Esteemed Brain Tonic and Intellectual Beverage, a cure for all nervous affections, sick headache, neuralgia, hysteria, and melancholy." *Dr. Pemberton* was proprietor of the Triplex Liver Pills and

Globe of Flower Cough Syrup Company. His new product contained cocaine, caffeine, and kola nut extract; it continued to include cocaine until 1903 when that ingredient was removed for racial reasons, but it was still called "dope" for many years to come. The name of this elixir is *Coca Cola.*

d.1892

William Bowman, English surgeon, anatomist, and ophthalmologist. He is remembered in *Bowman's glomerular capsule, Bowman's corneal membrane,* and *Bowman's glands of the olfactory mucous membrane.* He gave a classic account of striated muscle, proposed a theory of kidney function, and studied the actions of eye muscles and the lens. He was the first president of the Royal Ophthalmological Society.

March 30

b.1135

Abû Imrâm Mûsâ ibn Maimûn (Moses Maimonides) of Cordoba, Jewish physician, theologian, and philosopher. He wrote *Dalalat al-Ha'irir* or *A Guide for the Perplexed.* "Teach thy tongue to say 'I do not know'."

1842

Dr. Crawford W. Long (1815–1878), a simple and modest country practitioner, "discoverer of the use of sulphuric ether as an anesthetic in surgery on March 30, 1842, at Jefferson, Jackson County, Georgia," uses his discovery for the first time in an operation on his friend *James M. Venable.* He does not publish his results until 1849; *Morton's* "Letheon" comes out in 1846.

March 31

b. 1596

René Descartes, French mathematician, philosopher, anatomist, and physiologist. He studied medicine and law at Poitiers (1614–1616);

he founded analytic geometry and wrote on the anatomy of the eye and the mechanism of vision.

1777

Theodorick Bland is created a cavalry colonel in the Continental Army. Through his inept reconnaissance, this physician-descendent of *Pocahontas* contributes significantly to the American defeat at Brandywine. *'Light Horse Harry' Lee: "Colonel Bland* was noble, sensitive, honorable, and amiable; but never intended for the department of military intelligence."

d.1917

Emil Adolf von Behring, a German physician and immunologist. He founded the science of immunology and introduced the term antitoxin. Eventually he won the first Nobel Prize in medicine for his development of the diphtheria antitoxin.

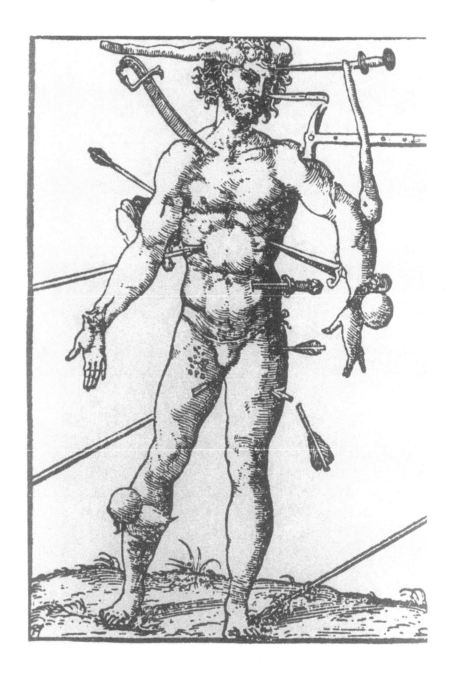

April

April is the cruelest month.

—*T. S. Eliot*

April 1

b.1657

William Harvey, English physician. "Ah! my old friend *Dr. Harvey*—I knew him right well—he made me sitt by him 2 or 3 hours together discoursing. Why! had he been stiffe, starcht, and retired, as other formall doctors are, he had known no more than they. From the meanest person, in some way, or other, the learnedst man may learn something. Pride has been one of the greatest stoppers of the advancement of learning." *John Aubrey,* from a play fragment appended to his *Brief Lives.*

William Harvey

1717

Lady Mary Wortley Montagu anticipates *Jenner* on vaccination: "I am patriot enough to take pains to bring this useful invention into fashion in England, and I should not fail to write to some of our Doctors very particularly about it if I knew any one of 'em that I thought had Virtue enough to destroy such a considerable branch of their Revenue for the good of Mankind, but that Distemper is too beneficial to them not to expose to all their Resentment the hardy wight that should undertake to put an end to it." (Letter to *Sarah Chiswell*) *Lytton Strachey:* "She was, like her age, cold and hard; she was infinitely unromantic; she was often cynical, and sometimes gross."

1792

Dominique Jean Larrey joins the Army of the Rhine. His *Memoirs of Military Surgery* and *Campaigns of the French Armies* remain classic Napoleonic documents.

April 2

b.1669

Jacob Benignus Winslow, Dutch anatomist. The peritoneal foramen epiploicum is named after him.

1893

New York physician **Robert Buchanan** is found guilty of murdering his wife Annie with morphine; he used atropine to conceal the pinpoint pupils. Defense lawyer and former physician **William O'Sullivan** with the assistance of **Dr. Victor Vaughan** destroyed the prosecution's toxicology case, but the accused was, unfortunately for him, allowed to testify in his own behalf.

d.1922

Hermann Rorschach, Swiss psychiatrist. He developed a test to assess intelligence and emotional elements of the personality based on an interpretation of inkblots (1921).

April 3

1537

François Rabelais takes his degree of licentiate in medicine at the University of Montpellier. "Science without conscience is the death of the soul."

April 4

d.1763

Johann Georg Roederer, German obstetrician and astronomer. He founded the first German obstetrical clinic, studied the fetal circulation, the position of the fetus, and the mechanism of delivery. He wrote *Elementia artis obstetricae* (1753).

d.1774

Oliver Goldsmith, English writer and physician. *"Goldsmith's* mind was entirely unfurnished. When he was engaged in a work, he had all his knowledge to find, which when he found, he knew how to use, but forgot it immediately after he had used it." *(Joshua Reynolds)* It is thought that he hastened his own end by treating himself. When asked on his deathbed if his mind was at ease, he replied, "No, it is not." "No man was more foolish when he had not a pen in his hand, or more wise when he had." *(Dr. Johnson)*

1783

The London newspapers requested former students of *Dr. William Hunter* to donate 1 guinea toward his funeral expenses since his fortune was tied up in his museum.

1969

The world's first totally artificial heart (Dacron and plastic) is implanted by *Dr. Michael DeBakey,* at Baylor University, Houston, Texas. The patient survived four days.

April 5

b.1827

Joseph Lister, Scottish surgeon and discoverer of the antiseptic principle in surgery. "Subtract his professional and scientific activities from his life and literally there is nothing of any moment to record." (*Kenneth Walker*) His father, *Joseph Jackson Lister* (1786–1869), had developed the achromatic lens of the modern microscope (1830). He is eponymously remembered in two bacterial genera, *Listerella* and *Listeria,* and a mouthwash.

1893

Robert Louis Stevenson writes to *Conan Doyle*, "....can this be *our old friend Joe Bell?*" He was referrring to the Edinburgh physician who served as one of the models for Sherlock Holmes.

b.1899

Alfred Blalock, American surgeon who in 1944 helped *Helen Taussig* perform the first "blue baby" operation.

April 6

1726

First published use of obstetrical forceps by *William Gifford*, surgeon and man-midwife.

d.1759

Johann Gottfried Zinn, German physician, botanist, and anatomist. His *Anatomy of the Eye* was the first book on the subject, and many ocular structures are named for him, as is the flower zinnia.

b.1886

Walter Edward Dandy, Missouri neurosurgeon who invented ventriculography while a chief resident. *Dandy-Walker syndrome* is hydrocephalus secondary to atresia of the foramen of Magendie.

b.1902

Meyer Perlstein, child neurologist and founder of the American Academy of Cerebral Palsy. "Pediatricians eat because children don't." "If your time hasn't come, not even a doctor can kill you."

April 7

d.1589

Julio Caesar Aranzio, Italian physician, surgeon, and anatomist. A pupil of *Vesalius*, his *De humano foetu opusculum* (1564) described the ductus arteriosus and the ductus venosus (the ductus venosus arantii). He is credited with recording the first case of pelvic deformity as well as an early description of a median cleft between the two lamina of the septum pellucidum (the *ventricle of Arantius*, fifth ventricle, pseudocele). The *bodies of Arantius* are tubercles in the semilunar valves of the aorta.

d.1789

Pieter Camper, Dutch physician, anatomist, naturalist, paleontologist, and artist. He was an outstanding teacher who discovered that the bones of birds contained air and described the fibrous structure of the crystalline lens. Among several eponyms, *Camper's facial angle* is a measure of prognathism.

1853

John Snow places Queen Victoria under chloroform anesthesia for the delivery of her fourth son, *Prince Leopold*, Duke of Albany. *Dr. Snow* became the first London specialist in anesthesia.

April 8

b.1695

Johann Christian Günther, dissolute German poet who studied medicine at his family's insistence. *Goethe* said of him: "He had everything needed to create a second life in life through poetry—except character."

b.1817

Charles-Edouard Brown-Séquard, a founder of modern endocrinology who pioneered in using organ extracts for replacement therapy. *Brown-Séquard syndrome* is a hemiplegia with sensory loss on the opposite side that occurs in a spinal cord hemitransection (1850).

b.1850

William Welch, American bacteriologist. "I plead for a little more of the cultural side in medicine." *Clostridium welchii* is named after him.

b.1869

Harvey W. Cushing, pioneer American neurosurgeon, neuroendocrinologist, and medical illustrator. He studied the regulatory functions of the pituitary and hypothalamus; in *Cushing's disease*, a pituitary tumor produces truncal ("buffalo hump") obesity, osteoporosis, purple striations, and abnormalities of carbohydrate metabolism. He wrote *Consecratio Medici* and a Pulitzer Prize winning biography, *The Life of Sir William Osler.*

April 9

d.1553

François Rabelais, French physician and anatomist, Renaissance writer. "*Rabelais*, you may remember, when physician to the Hotel Dieu in Lyons, published almanacs for the years 1533, 1535, 1541, 1546. In the title page he called himself 'Doctor of Medicine and Professor of Astrology,' and they continued to be printed under his name until 1556." (*Osler*) His will read, "I have nothing. I owe much. I leave the rest to the poor."

François Rabelais

b.1802

Elias Lönnrot, Finnish philologist, folklorist, and general practitioner. He compiled the *Kalevala* (1836), the Finnish national epic, while making house calls.

April 10

1697

Dated preface to *John Browne's Myographia Nova: or, A Graphical Description of All the Muscles in Humane Body, as they arise in Dissection.*

b.1707

John Pringle, Scottish physician. He originated the idea for the Red Cross, popularized the term influenza, and was one of the founders of modern military and sanitary medicine. He wrote *Observations on the Diseases of the Army* (1752).

John Browne

b.1755

Christian Friedrich Samuel Hahnemann, German founder of homeopathy. This system of medicine is based on the principle of *similia similibus*—many diseases can be cured by the administration of very small amounts of drugs that produce a condition similar to the disease. More traditional medicine is described as allopathy and is based on the principle of *contraria contrariis.*

April 11

b.1755

James Parkinson, an English surgeon, geologist, paleontologist, political agitator, and reformer. His 1817 *Essay on the Shaking Palsy* described paralysis agitans or *Parkinson's disease.* He was the first to recognize perforation as a cause of death in appendicitis (1812). A founder of the Geological Society of London, he wrote the classic paleontology text *Organic Remains of a Former World* (3 vols).

April 12

1829

Dr. Jules Cloquet amputates a breast from a woman asleep under hypnosis.

b.1853

James Mackenzie, Scottish cardiologist, student of arrhythmias. "The seeming exactness of a mechanical device appeals much more strongly to certain minds than a process of reasoning."

1856

Artificial respiration is described by *Dr. Marshall Hall* (1790–1857) in the *Lancet. Dr. Hall,* an English physician and neurophysiologist, also described *Hall's facies,* the disproportion between the forehead and the face in hydrocephalus.

Dr. Marshall Hall

1955

The outcome of the *Salk* vaccine trials is announced by the National Foundation of Infantile Paralysis at the University of Michigan. Vaccination of almost half a million children resulted in several hundred vaccine-related cases of polio, 150 paralyzed children, and 11 deaths.

April 13

b.1760

Thomas Beddoes, English physician and scientific writer. He wrote the classic *Observations on the Nature of Demonstrative Evidence* (1793), a mathematical treatise.

1872

An American general practitioner, *George Huntington* (1850–1916), at the age of 22 published a concise description of hereditary chorea in the *Medical and Surgical Reporter of Philadelphia.*

April 14

1648

Thomas Sydenham is created a Bachelor of Medicine. A practical and sceptical physician, "*Sydenham* was called 'a man of many doubts' and therein lay the secret of his great strength....*Sydenham* broke with authority and went with nature....He laid down the fundamental proposition, and acted upon it, that 'all disease could be described as natural history.'" *(Osler) Sydenham's chorea* is the post-rheumatic fever of *Saint Vitus's dance.*

b.1787

Pierre Charles Alexandre Louis, French physician. His greatest influence was as a teacher, especially with his American students. He pioneered in the Numerical Method, the application of basic statistics to clinical observations. He was the first to use a watch to time the pulse. "My name is Pulse-feel: A poor doctor of physick." *(Brome, City Wit)*

April 15

b.1492

Leonardo da Vinci, Italian Renaissance artist who made many contributions to anatomy. Around 1509 he acquired a copy of *Galen's De usu partium* and he became a confirmed Galenist. *William Hunter,* on reviewing his anatomical drawings, said, "*Leonardo* was the best Anatomist at that time in the world."

Leonardo da Vinci

b.1641

Robert Sibald, Scottish physician, botanist, and historian. The plant genus *Sibbaldia* is named for him.

d.1791

Alexander Garden, Scottish physician and naturalist. The gardenia is named for him.

1821

"15. To Surgeon-in-Chief *Larrey*, 100,000 francs; he is the most virtuous man I have ever known." (Item in *Napoleon's* will)

April 16

1639

License to print is granted to the *Regimento* of *Guilherme Escoph de Esens*, German chief surgeon of the Galleys of the Kingdom of Portugal.

April 17

1616

In the second Lumleian Lecture at the New Anatomical Theatre of the Physician's College, Amen Street, *William Harvey* presented his then novel ideas on the circulation of the blood.

b.1627

Henry Vaughan, Welsh physician–naturalist, religious–metaphysical poet. He had studied law, fought on the Royalist side in the Civil War, and then became a doctor. After publishing his *Silex Scintillans* (1650), he became known as the "Silurist." "They are all gone into the world of light." *Siegfried Sassoon* wrote of him:

> Through pastures of the spirit washed with dew
> And starlit with eternities unknown.
> Here sleeps the Silurist; the loved physician;
> The face that left no portraiture behind;
> The skull that housed white angels and had vision
> Of daybreak through the gateways of the mind.

---- *April* ----

1649

"I fell dangerously ill of my head; was blistered and let blood behind the ears and forehead." *John Evelyn, Diary.*

1917

Constantin Baron Economo von San Serff (1876–1931) first describes *encephalitis lethargica.*

April 18

d.1751

Laurent Garcin, French surgeon and botanist. He was honored by *Linnaeus* in the name of the tree species, *Garcinia.*

1767

Karl Pieter Thunberg, Swedish physician, botanist, and explorer, receives his medical degree at Uppsala. He wrote *Flora Japonica* (1784), *Fauna Japonica* (1823), and *Florula Javanica* (1825).

1775

Joseph Warren, successful Boston physician and political activist, sends *Paul Revere* and *William Dawes* on their famous midnight ride; he then took an active part in the next day's fighting. He later declined the post of Physician General and was elected Major General of Militia. He was killed at Bunker Hill.

d.1831

John Abernethy, English surgeon, pupil, and brusque successor to *John Hunter.* He wrote *Lectures on Anatomy, Surgery and Pathology* (1828); *Abernethy's carcinoma* is a malignant liposarcoma of the trunk, and *Abernethy's fascia* covers the external iliac artery. His advice to a mother with a too tightly laced daughter was, "Why, Madam, do you know that there are upward of thirty yards of bowels squeezed underneath that girdle of your daughter's? Go home and cut it; let Nature have fair play, and you will have no need of my advice." "However he might sometimes forget the courtesy due to his private patients, he was never unkind to those whom charity had confided to his care." (*Macilwain*) **Lombroso's *The Man of Genius*** records the following exchange:

> *Abernethy:*   You need amusement; go and hear *Grimaldi.*
>   He will make you laugh, and that
>   will be better for you than any drugs.
>
> *Patient:*   My God, I *am Grimaldi!*

He was nicknamed "Dr. My-book" because of his habit of telling his patients "Read my book" whenever any medical questions arose.

April 19

1775

Dr. Samuel Prescott's (1751–1777) courting of *Miss Milliken* in Lexington is interrupted on the night of the eighteenth when he is invited to join **Paul Revere** and **William Dawes**; the latter are captured and he becomes the only messenger to reach Concord. He later dies as a prisoner of war.

d.1876

Sir William Robert Willis Wilde, Irish ophthalmologist and otologist. He was an antiquarian, the founder of the *Dublin Quarterly Journal of Medicine,* and the father of **Oscar Wilde.**

1910

The first successful treatment of syphilis with **Paul Ehrlich's** "magic bullet," compound "606."

1945

The French court issues a warrant for the arrest of the physician and misanthropic writer **Louis-Ferdinand Céline** for collaboration with the Vichy government.

April 20

1775

The Medical Department of the American Army is formed the day after the battle of Lexington.

April 21

b.1634

Jan van Riebeck, Dutch naval surgeon, founder of Cape Town.

1892

Inscription on *Dr. Hunter Robb's* copy of the first edition of *Osler's The Principles and Practice of Medicine.* "N.B. This book was conceived in robbery and brought forth in fraud. In the spring of 1891 *I coolly* entered in & took possession of the working room of *Dr. Hunter Robb*—popularly known as the *Robin.* As in the old story of the Cookoo & the hedge sparrow I just turned him out of his comfortable nest, besplattered his floor with pamphlets papers and trash & played the devil generally with his comfort....Signed on behalf of the Author 4/21/92 E.Y.D."

1908

The medical missionary *Wilfred Grenfell* (1865–1940) falls through the ice and is saved by the skins of three faithful sled dogs whose sacrifice is remembered in a monument he later erected on the site. He penned the autobiographical *Forty Years for Labrador.*

1913

William Osler delivers the Silliman Lectures, published eight years later as *The Evolution of Modern Medicine.*

April 22

d.1607

Girolamo Rossi (*Rebeus, DeRubeis*), Italian physician and historian.

1661

"In dissecting in my first study, the head of a sheep which I had purchased on the 7th of April, I discovered a canal, which so far as I know, no anatomist has as yet described." Letter of *Niels Stensen* to *Thomas Bartholin.*

b.1898

William Stewart Duke-Elder, Scottish ophthalmologist, author of the classic *Textbook of Ophthalmology.*

April 23

d.18m95

Carl Ludwig, German physiologist who contributed to the study of the kidney and the circulation. He was "the greatest teacher of physiology who ever lived."

d.1931

Francis Xavier Dercum, Philadelphia neurologist. He collaborated with *Edward Muybridge* on the latter's *Animal Locomotion.* They produced the first photographs of normal and pathological gaits and of an artificially induced seizure. *Dercum's disease* is *adiposa dolorosa.*

Carl Ludwig

April 24

1678

Sir George Wakeman, Catholic physician to *Queen Katherine*, the wife of *Charles II,* is charged with high treason in plotting to murder the king. The infamous *Titus Oates* testifies to witnessing the plot at a Jesuitical convention on the 24th of April.

1769

George Armstrong opens the first dispensary for the infant poor. He is credited with one of the earliest descriptions of pyloric stenosis.

d.1964

Gerhard Domagk, German biochemist. He won the Nobel Prize for his work in developing the sulfonamides, but the award was declined because of the Nazi regime.

April 25

d.1549

Andrew Boorde, English physician. He was a Carthusian bishop who wrote the first printed handbook of Europe, published the first specimen of the Gypsy language, and is credited with the earliest modern work on hygiene. His books include *Dyetary* (1542), *Boke of Berdes,* and *Breyary of Health* (1547). He was once claimed as the original "Merry Andrew."

d.1728

John Woodward, English physician and geologist. He was the first to recognize strata in the earth's crust.

1763

Reverend Edmund Stone reports to the Royal Society the effectiveness of willow bark in the treatment of rheumatism. The active agent will later be found to be salicin, a substance similar to the basic component of aspirin.

1777

A Royal Declaration establishes the College de Pharmacie; the 'apothicaire' becomes the 'pharmacien.'

William Beaumont

1792

The first use of the guillotine, a medically humane invention by the French surgeon, *Dr. Antoine Louis* and the German harpsichordist, *Tobias Schmidt.*

d.1853

William Beaumont, American backwoods physiologist.

April 26

d.1558

Jean François Fernel, French physician, astronomer, and mathematician, the French *Galen.* He described appendicitis (the "iliac passion"), peristalsis, endocarditis, the central canal of the spinal cord, and intestinal obstruction. He was the first to use the terms "physiology" and "pathology" in their modern sense.

Jean Fernel

1610

Jeremiah Trautman performs the first fully documented Cesarean section in Wittenberg. The mother lived 25 days, the child 9 years.

b.1829

Christian Albert Theodore Billroth, German surgeon and anatomist. He was the father of modern intestinal surgery. *Billroth's cords* refer to the arrangement of sex cells in the developing ovary. "It is quite correct to distinguish between medical science and the physician's art."

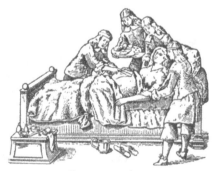

Cesarean section

April 27

1527

Publication of *The Vertuose Boke of Distyllacyon of the Waters of all manner of Herbes* by **Laurens Andrewe**.

1654

The founding of the **Salpétrière**, a general asylum for the impecunious of Paris. Its name came from the fact that it was located on a site where salt peter had been prepared.

1887

George Thomas Morton, son of the pioneer of anesthesia, performs the first appendectomy.

April 28

b.1828

Leopold Auerbach, German physician and anatomist. Auerbach's ganglia and plexus refer to the autonomic fibers between the muscular coats of the intestine.

d.1842

Charles Bell, Scottish physician. He described Bell's facial palsy; the *Bell-Magendie law* located motor functions in the anterior spinal roots and the sensory functions in the posterior roots. Two of his most famous books were *The Anatomy of Expression* and the teleological classic, *The Hand, Its Mechanism and Vital Endowments as Evincing Design*.

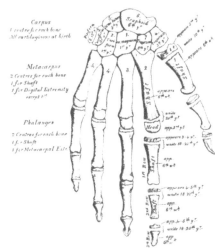

From *Gray's Anatomy*

April 29

d.1833

William Babington, Irish physician and mineralogist. The mineral Babingtonite is named for him.

1937

Fuller Albright describes a syndrome of *osteitis fibrosa disseminata*, which entails pigmentary abnormalities and endocrine dysfunction, including precocious puberty in females.

April 30

Walpurgis Night, the eve of May Day. *Saint Walpurga* (d.779), an English princess, studied medicine and founded a monastery at Heidenheim in Germany. Pictures show her with a flask of urine in one hand and bandages in the other.

d.1934

William Welch, American bacteriologist.

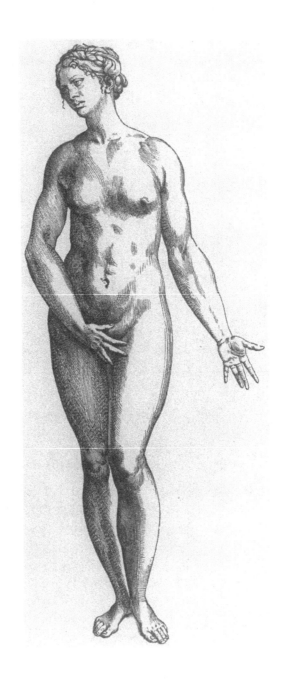

May

May 1

b.1785–d.1856

George James Guthrie, British surgeon who served in the Peninsula campaign. *Guthrie's muscle* is the deep transversus perinei; he devised several new procedures and accomplished the first successful ligation of the peroneal artery on the battlefield of Waterloo.

1816

René Laënnec, a French physician, invents the stethoscope. Called to examine an obese young woman with symptoms of heart disease, he found direct auscultation impermissible because of the patient's age and sex. An accomplished flutist, he rolled up a cylinder of paper and discovered that he could hear the heart sounds better than usual. His *De l'Auscultation Médiate* (1819) remains "one of the eight or ten greatest contributions to the science of medicine." *(Osler)* In 1800 he published *La Guerre des Venetes: Poeme Epique Hero-Comique.*

Santiago Ramón y Cajal

b.1852

Santiago Ramón y Cajal, Spanish neuroanatomist and neurohistologist. An erratic and rebellious student, he was apprenticed first to a barber and then to a cobbler. In addition to his technical writings, he wrote *Coffee-House Chatter* and *The World as Seen at Eighty.*

83

May 2

b.1886

Gottfried Benn, German physician and expressionist poet whose work reflects a mood of isolation and nihilism.

b.1903

Benjamin Spock, American pediatrician and political activist. He wrote the popular book, *Baby and Child Care*.

May 3

1140

Judah Halevi (c. 1086–1141) Spanish Jewish physician poet, liturgist, and enigmatologist enters Egypt at Alexandria. The author of *Al-Khazari* and *Diwan* will die within the year.

1765

Drs. John Morgan and *William Shippen, Jr.*, established a Medical Department in the College of Philadelphia as the first medical school in the American colonies; this would later become the University of Pennsylvania Medical School. These two doctors would fall out during the American Revolution and would spend the rest of their days feuding and quarreling.

1957

Male nurse *Kenneth Barlow* murders his wife with an injection of insulin.

May 4

b.1825

Thomas Henry Huxley, evolutionary biologist, "Darwin's bulldog." *Huxley's layer* is the inner layer of the root sheath of the hair follicle. "*Huxley* found controversy the spice of life." (*Cannon*)

1876

Dr. Adam Hammer of Saint Louis correctly diagnoses coronary occlusion during life for the first time. The case is published February 2, 1878.

1892

Arthur Conan Doyle writes to *Joseph Bell* (1837–1911), consulting surgeon to the Royal Infirmary and Royal Hospital for Sick Children, and identifies him as the primary model for Sherlock Holmes. *Bell* was the last in a long family line of physicians and had been one of *Doyle's* teachers at medical school.

May 5

1816

The first published poem by *John Keats* appears in *The Examiner.*

> *Yet the sweet converse of an innocent mind,*
> *Whose words are in ages of thoughts refined*
> *Is my soul's pleasure....*
> —*O Solitude*

1846

The American Medical Association is proposed in convention. It will be established one year later.

1881

Louis Pasteur begins his public demonstration of the effectiveness of vaccination for sheep anthrax.

May 6

b.1830

Abraham Jacobi, American pediatrician. He was one of the founders of the *American Journal of Obstetrics* and president of the American Medical Association. He wrote books on dentition, infant feeding, diphtheria, and intestinal diseases of children.

b.1856

Sigmund Freud, Viennese neurologist who pioneered in the classification of the cerebral palsies of children, the use of cocaine as an anesthetic, and the attribution of all the major ills of civilization to sexual repression. On himself: "I am not really a man of science, not an observer, not an experimenter, and not a thinker. I am nothing but by temperament...an adventurer...with the curiosity, the boldness, and the tenacity that belongs to that type of being."

1954

English medical student *Roger Bannister* runs the mile in 3:59.4. He will later succeed *Lord Brain* as the preeminent British neurologist.

May 7

b.1841

Gustave Le Bon, French physician and founder of social psychology. His seminal study, *La Psychologie des foules* (1895), was concerned with mob psychology. "One does not behave according to one's intelligence but according to one's character."

1889

Johns Hopkins Hospital opens.

1918

At a meeting of the Natural History and Medical Society in Heidelberg, Professor *Ernst Moro* discusses the first three months of life and at the end of his presentation mentions a peculiar infantile patterned response observed when striking the pillow.

May 8

d.1683

Peter Chamberlen, inventor of the obstetrical forceps, which were kept a family secret for almost a century. He advocated incorporating midwives and published many political and theological schemes. The Chamberlin family also marketed the Anodyne Necklace; this was used for teething and was composed from the bones of *Saint Hugh.* This saint had had a way with children, and when he would dip his fingers in holy water and rub their gums, their toothache would be cured. The efficacy of the necklace depended on this curative power having been transferred to his bones.

d.1969

Sydney Smith, British forensic pathologist, author of *Mostly Murder* (1960).

May 9

1738

John Wolcot is baptized. He would later become official physician to Jamaica, an ordained minister, and a painter. As *"Peter Pindar"* he would scurrilously lampoon and caricaturize in verse *George III, James Boswell,* and *Benjamin West,* among others.

d.1805

Friedrich von Schiller, regimental physician, poet, dramatist, historian, and literary theorist. Among his most famous works are *Don Carlos* (1787), *Wallenstein* (1799), and *Mary Stuart* (1800). His ode, *An die Freude,* was set to music by *Beethoven* as the choral finale to the Ninth Symphony.

1816

Jacques Delpech first attempts to cure clubfoot by a subcutaneous Achilles tenotomy.

May 10

d.1566

Leonard Fuchs, German physician and botanist. His 1542 herbal manual, *Historia Stirpium,* was a landmark in natural history because of its beautiful engravings. He is remembered in the name given to the genus of plants in the evening primrose family, "ladies' eardrops" or Fuchsia.

b.1888

Rudolf Schindler; in 1932 this gastroenterologist invented the flexible gastroscope.

1906

August von Wasserman (1866–1925), a German bacteriologist, describes a serodiagnostic test for syphilis.

b.1910

Eric Lennard Berne, American psychiatrist who invented and popularized transactional analysis in *Games People Play* (1964).

1929

Alexander Fleming publishes his report on the antibacterial effects of the penicillin mold.

May 11

d.1684

Daniel Whistler, English physician who had given one of the earliest descriptions of rickets and had been president of the College of Physicians. When he died in debt, his friends whisked his body away lest it be sold to be anatomized, with the price then applied against his debts.

May 12

National Hospital Day, so selected for *Florence Nightingale's* birthday.

May

1857

The *New York Infirmary for Women and Children* opens on *Florence Nightingale's* birthday under the direction of *Drs. Elizabeth* and *Emily Blackwell* and *Dr. Marie Zakzewska*, the first hospital to be operated exclusively by women.

May 13

b.1588

Ole Worm, Danish anatomist and theologian. He was Professor of Greek, Anatomy, and Philosophy in Copenhagen and brother-in-law to Caspar Bartholin (primus) whom he succeeded. The ossa suturalia are named for him—*Wormian bones.*

d.1742

Nicholas Andry, French physician who completed his classic *Orthopaedia* at the age of 83, the year before his death.

Treating wry neck, from Orthopaedia

b.1851

Spencer F. Baird first describes *Sylvicola kirtlandii* or Kirtland's warbler, named to honor *Jared Potter Kirtland* (1793–1877), American physician, teacher, horticulturist, and naturalist on whose estate the bird was captured.

Jared Kirtland

b.1857

Ronald Ross, British army surgeon. He was awarded the 1902 Nobel Prize for discovering that the Anopheles mosquito transmits malaria.

b.1883

George Papanicolaou, Greek-American physician who develops the Pap smear for uterine cancer in 1933.

May 14

1796

Edward Jenner vaccinates eight-year-old *James Phipps* against smallpox using virus from a cowpox pustule from *Sarah Nelmes*, a dairymaid. He publishes his results in *An Inquiry into the Causes and Effects of the Variola Vaccine* (1798).

May 15

Feast of *Saint Dymphna*, patron saint of the insane, the emotionally disordered, the psychotic, and the mentally retarded. It was under her patronage that, in the late Middle Ages in Gheel, Belgium, a tradition of very humane, community-based treatment for the mentally disabled was started.

d.1482

Paolo dal Pozza Toscanelli, Florentine physician and mapmaker. He dabbled in astronomy and plotted the course of *Halley's* comet in 1456. His cartography exerted a strong positive influence on *Christopher Columbus.*

1672

Dated introduction of *Hugh Chamberlen's* translation of *François Mauriceau's The Accomplisht Midwife.* Hugh tried to sell the family secret (obstetrical forceps) in Paris, but an unfortunate fatal outcome with a patient of *Mauriceau* contributed to a longer delay in the forceps coming into general use.

1724

'Déclaration royale' recognizes the apothecaries' right to visit the patient's bedside in the absence of a doctor.

1847

"As of today, May 15, 1847, every doctor or student who comes from the dissecting room is required, before entering the maternity wards, to wash his hands thoroughly in a basin of chlorine water which is being placed at the entrance. This order applies to all, without exception." *I. P. Semmelweis'* sign posted outside the lying-in section of the Vienna General Hospital.

d.1851

Samuel George Morton, American physician and naturalist. He described the fossils brought back on the Lewis and Clark expedition, studied the comparative anatomy of the human skull and identified a new species of hippopotamus.

b.1862

Arthur Schnitzler, Austrian physician, dramatist, and novelist. He wrote on love and sex from a *Freudian* perspective; his most famous work was *Reigen* (1903).

b.1891

Mikhail Afanasievich Bulgakov, Russian physician, venereologist, and novelist. He wrote the diabolical *The Master and Margarita*, the surrealistic *The Heart of a Dog*, and the realistic *A Country Doctor's Notebook*. "You are a very good doctor, but you've chosen the wrong career. You should have been a writer."

May 16

Feast of *Saint Peregrine* (d.261), bishop and martyr. He is the patron saint of those suffering cancer, leg ailments, incurable diseases, and cancer of the leg.

1769

Dr. Samuel Bard delivers the commencement address at King's College, "A discourse upon the duties of a physician, with some sentiments on the usefulness and necessity of a public hospital." When published in an 18-page pamphlet it will become the first book on medical ethics.

b.1845

Elias Metchnikoff, Russian pathologist who developed the theory of phagocytosis and the concept of inflammation as a defense reaction. He shared the 1908 Nobel Prize with *Paul Ehrlich.*

d.1882

Thomas Wakley, English physician, medical reformer, and MP. He improved inquest procedures and reformed the office of coroner; in 1823 he started the *Lancet.*

1921

Frederick Banting and *Charles Best* ligate the pancreatic ducts of their first experimental dog, leading to the extraction of "isletin," later called insulin.

May 17

b.1749

Edward Jenner, English physician who studied anatomy and surgery under *John Hunter.* The latter gave him this famous advice, "Don't think, try!"

d.1801

William Heberden, English physician and classical scholar. He described angina pectoris, hypochondria and hysteria, arthritic nodules, and distinguished chickenpox from smallpox. "New medicines, and new methods of cure, always work miracles for a while."

d.1809

Leopold Auenbrugger, Austrian physician. His *Inventum novum ex percussione thoracis humani* inaugurated the art of physical diagnosis with his discovery of percussion. He was an accomplished musician and he had watched his innkeeper father tap on barrels to assess their fullness, and these experiences undoubtedly contributed to his insight.

May 18

b.1795

Alfred Marie Velpeau, French surgeon, anatomist, and textbook writer. The bandage for a fractured clavicle is named for him, as are a number of anatomical structures. He wrote a textbook on the art of delivery as well as a pioneering work on the diseases of the female breast, *Traité des maladies du sein et de la region mammaire* (1842).

d.1861

Friedrich August von Ammon, German ophthalmologist. *Ammon's scleral prominence* is located in the posterior aspect of the globe of the eye of the fetus at the third month. He introduced both blepharoplasty and dacrocystotomy.

May 19

d.1586

Adam Lonitzer (Lonicerus), German physician, botanist, and zoologist, author of *Kreuterbuch.* The fly honeysuckle is named for him— *Lonicera canadensis.*

d.1795

Josiah Bartlett, physician, medical reformer, Revolutionary patriot, chief justice, and governor of New Hampshire.

---------------------------------- *May* ----------------------------------

b.1857

John Jacob Abel, Baltimore pharmacologist and physiologist. He pioneered in the study of endocrine secretions, isolated epinephrine, developed plasmapheresis, obtained crystalline insulin, discovered that phenolsulfonphthalein was excreted by the kidney, developed a kidney function test based on that observation, and was instrumental in the use of phenoltetrachlorthalein as a test for liver function. He also studied the use of the phthaleins as purgatives.

May 20

b.1537

Hieronymus Fabricius (Gerolamo Fabrizio d'Acquapendente), Italian nobleman, surgeon, anatomist, and embryologist. He succeeded *Fallopius* and was a teacher of *William Harvey.* He wrote *De formato foetu* (1604), and his *De venarum otiolis* (1574) described "little doors" or valves in the veins.

1747

James Lind begins his classic experiments on the treatment of scurvy aboard the HMS Salisbury.

1862

Dr. Benjamin Franklin Goodrich is contracted as a surgeon with the Army of the Potomac. After the Civil War, B.F. Goodrich will start a rubber company in Akron, Ohio.

May 21

b.1860

Willem Einthoven, Dutch physician. He received the 1924 Nobel Prize in Physiology for the development of the electrocardiogram.

1862

US Surgeon General *William Alexander Hammond* (1828–1900) announces his intention to establish an Army Medical Museum; in 1949 this becomes the Armed Forces Institute of Pathology (AFIP).

1876

Italian surgeon *Edoardo Porro* performs the first modern Cesarean section on 25-year-old *Julie Covallini. Porro's operation* is a Cesarean section with a hysterectomy.

May 22

1537

François Rabelais takes his degree in medicine at the University of Montpellier. His Pantagruelism, "a certain gaiety of spirit, pickled in disdain for fortuitous things," was a means to make the sick better. He regarded the cheerful bearing of the physician of the greatest importance to the patient's improvement.

d.1819

Hugh Williamson, American statesman and scientist, physician, patriot, and mathematician. He collaborated with *Benjamin Franklin* in the latter's experiments on electricity and studied comets and climate.

b.1859

Arthur Conan Doyle, physician creator of Sherlock Holmes, Brigadier Gerard, Sir Nigel Loring, and Professor George Edward Challenger.

May 23

b.1718

William Hunter, Scottish anatomist and surgeon, "the first of the Hunters." He was Professor of Anatomy at the Royal Academy of Arts and wrote *The Anatomy of the Human Gravid Uterus* (Baskerville Press, 1774). He "worked till he dropped and lectured when he was dying." (*Stephen Paget*)

b.1734

Franz Anton Mesmer. Under the protection of *Marie Antoinette*, this Viennese physician received a grant of 30,000 francs from *Louis*

XVI to study the magnetic influence of the stars on human beings. His *Mémoire sur la découverte du magnétisme animal* (1779) described cures with magnets and hypnosis.

b.1926

John Hilton Knowles, iconoclastic director (its youngest) of the *Massachusetts General Hospital.* He criticized unnecessary surgery and high fees and advocated preventive medicine and national health insurance. "The AMA operating from a platform of negative vigilance presents no solutions, but busily fights each change and then loudly supports it against the next proposal."

May 24

b.1757

William Charles Wells, American Loyalist physician who returned to England. He authored a classic description of rheumatic heart disease, recognized rheumatic nodules on tendons, noted hematuria and albuminuria in dropsy, discovered that erysipelas is contagious, and described the concept and importance of the dew point.

1798

Philippe Pinel (1745–1826) cuts the chains from the madmen at the Bicêtre Hospital. This famed psychiatric nosologist wrote *Traité médico-philosophique sur l'aliénation mentale ou la manie* (1801).

b.1898

Helen Taussig, American pediatric cardiologist who designed the first operation to save "blue babies." She was the first woman to hold a full professorship at Johns Hopkins.

May 25

1706

Ukase of *Peter the Great* founds the first Russian Hospital in Moscow.

1753

Captain James Lind publishes his report on the prevention of scurvy by intake of citrus fruits, *Treatise of the Scurvy.* He also wrote *Essay on the Health of Seamen* (1757).

May 26

d.1569

Guido Guidi (Vidius), a Florentine nobleman and physician. His *Chirurgia e Graeco in Latinum conversa* was printed in 1544 at the home of his friend, *Benvenuto Cellini*; the illustrations were the work of *Primaticcio* (*Francesco Primaticcio* of Bologna, 1504–1570), painter, architect and Abbot of San Marco. The *Vidian nerve, canal, artery,* and *vein* are named for this anatomist.

1770

Oliver Goldsmith publishes "The Deserted Village."

b.1837

Henry Hicks, Welsh physician, specialist in mental diseases, and geologist who discovered the Middle Cambrian.

1917

W. Somerset Maugham marries *Syrie Wellcome.*

May 27

d.1315

Ibn Al-Raqqam, Spanish Moslem physician, mathematician, and astronomer.

1755

The cornerstone of the Pennsylvania Hospital is laid with *Benjamin Franklin's* inscription.

b.1894

Louis Ferdinand Céline (Henri Louis Destouches), French physician in the slums of Paris. He wrote *Journey to the End of Night* and *Death on the Installment Plan.* His fiction was hallucinatory, obscene, and absurd, reflecting morbid despair and black humor. After World War II he was jailed for collaboration with the Germans. "The medical is an invidious profession. When one's practice is among the rich one looks like a lackey; when it's among the poor like a thief."

d.1981

Kit Pedler, British ophthalmologist, cocreator of Cybermen, Cybermats, and UNIT; scientific advisor to *Doctor Who.*

May 28

b.1780

Nathaniel Chapman, first president of the American Medical Association.

b.1916

Walker Percy, American physician, novelist, and philosopher. Author of *The Moviegoer, The Second Coming, Lancelot,* and *Lost in the Cosmos: The Last Self-Help Book.*

May 29

1653

Publication of *De Morbis Puerorum* or a *Treatise of the Diseases of Children* by **Robert Penell.** "Some it may be will be offended at what I have written because it is in the Mother-tongue."

May 30

1903

William Boog Leishman (1863–1926), British tropical pathologist, reports intracellular protozoan forms of the parasite causing kala-azar or Dumdum fever. These will be named *Leishman-Donovan bodies.*

d.1891

Benjamin Fordyce Barker, American obstetrician, first president of the American Gynecological Society. He wrote *Remarks on Puerperal Fever* (1857) and was the first American physician to use a hypodermic syringe.

d.1909

Désiré Magloire Bourneville, French physician who described a familial disorder including mental retardation with deterioration, epilepsy, and sebaceous adenomas of the skin. The disorder is known as *epiloia* or *Bourneville's disease.*

May 31

b.1819

William Worrall Mayo, founder of the Mayo dynasty.

1889

The Birth of Modern Endocrinology: *Dr. Charles-Edouard Brown-Séquard* reports increased vigor after injecting himself with animal testicular material.

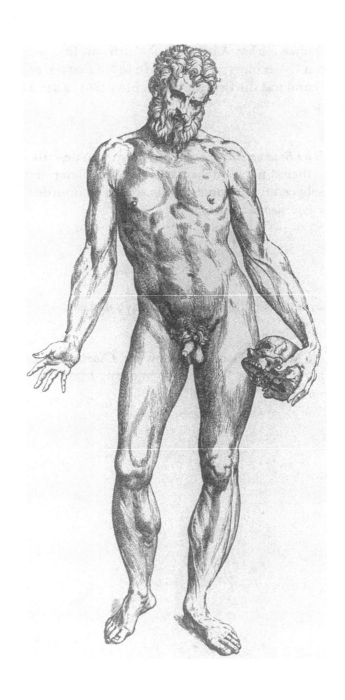

June

A noise like of a hidden brook
In the leafy month of June.

—*Samuel Taylor Coleridge*
The Rime of the AncientMariner

June 1

Festival honoring the Roman goddess *Carna* who protected and cared for the physical well-being of humankind; she was the "genius" of the flesh.

b.1822

Friedrich Matthias Claudius, Austrian anatomist who identified the polyhedral cells on the floor of the cochlear canal. He researched the hearing ability and apparatus of various animal species.

d.1941

Johannes "Hans" Berger, a German physician who devised a system of electrodes to measure brain wave patterns; he performed the first electroencephalogram (EEG) in humans in 1929.

June 2

Feast of *Saint Erasmus*. He was a bishop who died around the year 300. He became the patron saint of sailors, and his iconography included a windlass; this has been commonly misinterpreted as a disemboweling instrument, and he has thus been transformed into the patron saint of abdominal diseases. *St. Erasmus' disease* became intestinal colic.

St. Erasmus' intestines
being wound on a windlass

1697

Publication of *John Pechey's* A *General Treatise of the Diseases of Infants and Children.*

b.1824

Samuel Wilks, English physician who described bacterial endocarditis and alcoholic paraplegia. He gave the eponym to *Hodgkin's disease* after *Thomas Hodgkin* (1798–1866).

1881

Louis Pasteur demonstrates the first successful vaccination of sheep against anthrax.

June 3

d.1657

William Harvey, English physician. In his preface to *Walter Charleton's Chorea Gigantus: Or, The Most Famous Antiquity of Great Britain, Vulgarly Called Stone-henge, Standing on Salisbury-Plain, Restored to the Danes* (1663) *John Dryden* wrote

> The circling streams, once thought but pools, of blood
> (Whether life's fuel, or the body's food,)
> From dark oblivion Harvey's name shall save.

John Aubrey noted of *Harvey*, "He kept a pretty young wench to wait on him, which I guess he made use of for warmth-sake as *King David* did, and took care of her in his will."

1769

The transit of Venus over the sun's disc is observed and will become the subject of astronomical papers by *Hugh Williamson* (1735–1819), American astronomer, meteorologist, lawyer, and physician.

d.1836

Barry Edward O'Meara, Irish surgeon on the Bellerophon, physician to *Napoleon* in exile, author of *A Voice from St. Helena* (1823). On his return to England he was court-martialed and dismissed; he had been a double agent, taking money from both the Admiralty and *Napoleon*. The latter's journal said of him, "Le docteur n'est si bien pour moi que depuis que je lui donne mon argent."

b.1873

Otto Loewi, physician and neurophysiologist who won the 1936 Nobel Prize for work on the chemical transmission of nerve impulses.

June 4

b.1694

François Quesnay, French physician, surgeon, economist, and encyclopedist. He was a laissez-faire economist who held land to be the sole source of wealth.

1970

L-Dopa is approved by the Food and Drug Administration.

June 5

b.1757

Pierre Jean Georges Cabanis, French physician and philosopher; he was active in the revolutionary government and tried to rebuild French medicine on clinical grounds. His *Rapports du physique et du moral de l'homme* (1802) expressed a materialistic psychophysiology, and his *Les Révolutions et la réforme de la médicine* (1804) was an early history of medicine.

b.1932

Christy Brown, Irish novelist and poet who had severe cerebral palsy. "We slowly come to realize that there are people who understand,

people who have actually dedicated their lives towards helping us
and bringing us to a greater understanding of our own, so that in
the end something splendid is wrought out of our affliction." *My
Left Foot*

June 6

1822

Alexis St. Martin was wounded by a shotgun blast and treated by *Dr.
William Beaumont,* later author of *Experiments and Observations on
the Gastric Juice and the Physiology of Digestion* (1833).

b.1882

Kenneth MacFarlane Walker, British genitourinary surgeon who wrote
philosophical works and the autobiographical *I Talk of Dreams.*

d.1961

Carl Gustav Jung, Swiss psychiatrist, mythographer, and alchemist.

June 7

d.1728

Daniel Le Clerc, Swiss physician who is 1696 made the first attempt
at a synthetic survey of medical history.

b.1811

James Young Simpson, Scottish gynecologist who introduced chloroform anesthesia to obstetrics. Opinions of him varied from "a vulgar male midwife" to the "Napoleon of Scottish surgery." He
vigorously attacked *Lister's* antisepsis as worthless. One of his most
famous works was the 1847 tract *Answers to the Religious Objections
Advanced Against the Employment of Anaesthetic Agents in Midwifery
and Surgery* in which he quoted scripture against the idea that wom-
en had to labor in pain to bear children.

b.1870

Max Brödel, the founder of the Art of Medical Illustration in the United States. He described *Brödel's line* on the anterior surface of the kidney and devised *Brödel's stitch* for friable organs.

b.1873

Franz Weidenreich, German physician, anatomist, and anthropologist.

Osler's bookplate, by Brödel

June 9

b.1783

Benjamin Brodie, English surgeon, pioneer in vein surgery with the first surgical treatment of varicose veins. His conservative approach to the treatment of joint diseases reduced the number of amputations. He is remembered in a number of eponymous anatomical structures and pathological processes.

b.1836

Elizabeth Garrett Anderson, female medical pioneer and first woman mayor of Aldeburgh, Suffolk, her place of birth. "If one regards medicine as a pure science, her claims to distinction are not high. If one regards it as a skilled exercise in personal relationships, one must rate her very high indeed. It was as a general practitioner that she excelled." (*Naomi Mitchison*)

June 10

b.1735

John Morgan, Philadelphia physician, cofounder of the medical faculty at the University of Pennsylvania (1765).

June 11

b.1603–d.1665

Kenelm Digby, "an errant mountebank, the very Pliny of and the age for lying." He devised a sympathetic powder to heal wounds; the popularity of this weapon salve probably saved many lives for the simple reason that the concoction was applied to the offending instrument rather than the wound.

> *Under this stone the matchless Digby lies,*
> *Digby the great, the valiant and the wise:*
> *This age's wonder for his noble parts;*
> *Skill'd in six tongues, and learn'd in all the arts.*
> *Born on the day he died, th'eleventh of June,*
> *On which he bravely fought at Scanderoon.*
> *'Tis rare that one and self-same day should be*
> *His day of birth, of death, of victory.*

d.1738

Caspar Thomeson Bartholin, secundus, Danish physician and anatomist. Vaginal glands and the sublingual duct are named for him.

d.1864

Karl Gustav Jung, German anatomist. The musculus pyramidalis auriculae (or m. trago-helicinus) is known as *Jung's muscle.*

June 12

1667

Jean Baptiste Denis, professor of surgery at Paris and surgeon to *Louis XVI,* performs the first successful blood transfusion from a lamb to a dying man.

1888

First account of the surgical removal of a spinal cord tumor. The patient was *Captain Gilby,* the doctors *William Gowers* and *Victor Horsley.*

June 13

b.1773

Thomas Young, English physician, physicist, Egyptologist, and the Father of Physiological Optics. Known in Cambridge as "Phenomenon" Young, he described the phenomenon and cause of astigmatism, established the principle of interference of light, resurrected the wave theory of light, and proposed with *Helmholtz* a theory of color vision. He devised one of the first mathematical rules to determine medicine dosages for children. *Helmholtz* called him "one of the most clear-sighted men who ever lived" and he said of himself, "When I was a boy, I thought myself a man; now I am a man, I find myself a boy." His promise as a child prodigy was fulfilled when as an adult Egyptologist he helped decipher the *Rosetta Stone.*

d.1797

Simon Tissot, Swiss physician, a leading European authority on the harmful effects of masturbation.

b.1870

Jules Jean Baptiste Vincent Bordet, Belgian bacteriologist and Nobel Prize winner in Medicine. He discovered complement fixation, gave the first description of bacterial hemolysis, identified the bacillus of whooping cough, and a method of immunizing against it.

June 14

b.1868

Karl Landsteiner, Viennese pathologist. His discovery of blood group antigens permitted safe transfusions. In distinguishing blood types he also uncovered the Rh factor, and was given the 1930 Nobel Prize.

b.1884

Paul Budd Magnuson, American orthopedic surgeon. "I am very grateful that I was born...when they were willing to take candidates because they wanted to be doctors and the professors thought they might make good ones. They did not insist on a man of technical education or a straight 'A' record in school. They were looking for personality and interest and the brains to observe and absorb." *Ring the Night Bell*

b.1928

Ernesto "Che" Guevara, Argentinian physician, revolutionary, and guerilla.

June 15

Feast of *Saints Vitus* and *Modestus*. These are two of the five patron saints for chorea. *St. Vitus' dance* is frequently equated with *Sydenham's chorea*. *St. Vitus' dance of the throat* is stammering. (Vitus translates into English as Guy.) For understandable reasons, **St. Vitus** is also the patron saint of dancers and actors.

b.1809

Henri Roger, French physician who described the murmur of a ventricular septal defect in 1879. He was the first to give systematic clinical instruction in pediatrics in France and was never advanced academically beyond instructor.

1860

Fifteen students enter the first class at the Nightingale Training School for Nurses at St. Thomas's Hospital.

1867

The first successful gallstone operation (cholecystotomy) is performed on *Mary Wiggins* in McCordsville, Indiana, by the Hoosier doctor *John Stough Bobbs* of Indianapolis. The "Father of Cholecystotomy" thought he was operating on an ovarian tumor.

June 16

d.1858

John Snow, English physician who studied water-borne cholera epidemics and pioneered an apparatus to administer ether.

b.1905

Jakob Edward Schmidt, American physician and medicolegal lexicographer. He pioneered in the use of cesium as a catalyst for sewage oxidation and described the *Schmidt effect,* an optical illusion affecting vertical lines on a television screen.

b.1918

George C. Cotzias, neurologist who demonstrated that L-dopa reverses the symptoms of Parkinson's disease.

June 17

1836

The North American Academy of the Homeopathic Healing Art is chartered at Allentown, Pennsylvania. Founded by *Constantine Hering,* it offered the degree of Doctor of Homeopathia.

1867

The first operation is performed under antiseptic conditions. *Joseph Lister* amputates the cancerous breast of *Isabella Lister* (his sister) using carbolic acid as an antiseptic.

1950

The first kidney transplant is performed by *Dr. Richard Lawler* at the Little Company of Mary Hospital in Chicago.

d.1960

Harold Delf Gillies, English surgeon, the "Father of Plastic Surgery." "Often while lifting a face I have a feeling of guilt that I am merely making money. Yet is it not justified if it brings even a little extra happiness to a soul who needs it?"

June 18

d.1348

Gentile da Foligno, an Italian surgeon who wrote *Consilium* on the plague, in which work he described semina (seeds) of the disease and the reliquae of infectious materials left by patients. He also wrote on hydrotherapy. There was an astrological tinge to his medical writings. He eventually died of the plague.

June 19

Surgical instruments from Pompeii

Feast of *Saint Gervasius. St. Gervasius' sickness* is rheumatism.

b.1710

John Burton, English physician who served as the model for Dr. Slop in *Laurence Sterne's Tristam Shandy.* "Imagine to yourself a little, squat, uncourtly figure of a Doctor Slop, of about four feet and a half perpendicular height, with a breadth of back and a sesquipedality of belly, which might have done honour to a serjeant in the horse-guards." (II, 9)

June 20

1516

The *Grete Herball.* Imprented at London in Southwark by me Peter Treveris. MDXVI. The XX day of June.

d.1773

Georg Christian Füchsel, German geologist and physician. He was the first to recognize that rock layers and groups of related strata were geologic formations.

June 21

b.1759

James Ross, British military surgeon and linguist, member of the Royal Asiatic Society, first translator into English of Sadi's *Gulistan or Flower Garden* (1823). "The skill of the physician is advice." (III, iv)

1849

Marion Sims, pioneer American gynecologist, performs the first successful operation to treat vesicovaginal fistula in a long series of experiments on Negro slaves. He was "one of the most amiable and lovable of men."

b.1880

Arnold L. Gesell, American pediatrician and psychologist who pioneered in infant developmental assessment. He is the "Father of Developmental Pediatrics," and his work remains the basis for infant testing.

June 22

1874

"Finally...on June 22d, 1874, I flung to the breeze the banner of Osteopathy." *Andrew Taylor Still* (1828–1917), self-taught medical practitioner and founder of osteopathy, in his *Autobiography.*

June 23

d.1770

Mark Akenside, English poet and physician, author of *Pleasures of the Imagination.* He was notoriously ill-mannered and at one time ordered a patient who couldn't swallow his medicine to be removed from the hospital—"He shall not die in my hospital." He was satirized in *Tobias Smollett's The Adventures of Peregrine Pickle.*

June 24

Midsummer's Day or the Feast of *Saint John the Baptist,* the traditional day to pick therapeutic herbs.

1527

At the University of Basel, the "Luther of Medicine," *Paracelsus,* comes out holding in his hands *Avicenna's Canon,* the Bible of Medicine, and throws it into the flames. "Into St. John's fire so that all misfortune may go into the air with the smoke."

Paracelsus

1745

An act for making the surgeons and barbers of London two distinct and separate corporations. The principal actor behind this separation was the surgeon *William Cheselden* (1688–1752); he wrote *Anatomy of the Human Body* (1713) and *Osteographia* (1733), could perform a lithotomy in 54 seconds, and pioneered in the construction of an artificial pupil. The first master of the new company of surgeons was the rough and choleric *John Ranby,* nicknamed "The Blockhead."

1859

Swiss banker *Jean Henri Dunant* witnesses the horror of the Battle of Solferino between Austria and Napoleon III and decides to found the *International Red Cross.*

June 25

1663

Thomas Sydenham is admitted to membership in the College of Physicians of London. After an early career as a Puritan cavalry officer, *Sydenham* obtained his doctorate at 52 years of age. He founded the ontological concept of disease, in which each disorder is considered as a distinct entity. He was *"medicus in omne aevum nobilis."*

b.1824

Pierre Paul Broca, French surgeon and anthropologist, founder of the Society of Anthropologists in Paris (1859). He gave a pioneer description of aphasia, which he called aphemia, and the craniometric points for the inion, lambda, metropion, obelion, and opisthion; the third convolution of the left frontal lobe, the "speech area" of the cerebral cortex, is known as *Broca's area.*

1915

The *American College of Physicians* is founded.

June 26

d.1829

Gustave Henri Baron de Rosenthal (Lt. John Rose), Latvian-born physician, military surgeon, and duelist. He was "the only Russian on the American side in the War of Independence." He died Grand Marshal of Livonia.

d.1844

John Conrad Otto, American physician who wrote, "An Account of an Hemorrhagic Disposition Existing in Certain Families," the first description of hemophilia.

June 27

1721

Dr. Zabdiel Boylston (1680–1766) inoculates his own son during the Boston smallpox epidemic.

1831

Fridericus Wilhelmus Adamus Serturner (1783–1841), German pharmacologist, receives the Prix Montyon for his isolation from opium of the alkaloid morphine.

June 28

1774

Dr. J. Quier describes *Koplik's spots* during a measles epidemic in Jamaica (122 years before *Koplik*).

b.1873

Alexis Carrel, French surgeon who pioneered in blood vessel anastomoses and organ transplantations for which he received the 1912 Nobel Prize. He believed that black gowns were a necessary part of his surgical technique. In his classic *Man, The Unknown* he wrote, "The more eminent the specialist, the more dangerous he is."

June 29

d.1860

Thomas Addison, English physician who described both adrenal insufficiency and pernicious anemia. His 1849 *The Constitutional and Local Effects of Disease of the Suprarenal Capsule* is a classic.

b.1861

William J. Mayo, founder of the Mayo Clinic.

d.1881

Maurice Raynaud, French physician who described the eponymous *Raynaud's syndrome*, a paroxysmal bilateral cyanosis of the digits.

1888

The first appendectomy in England is performed by *Frederick Treves*.

June 30

b.1884

Georges Duhamel, French poet, novelist, playwright, and physician. After obtaining his medical degree in 1909, he divided his activities between scientific research and literary work.

June

1906

Pure Food and Drug Act enacted.

d.1961

Lee DeForest, American physicist who claimed to have invented the vacuum tube. In 1907 he designed an electrical high frequency surgical scalpel.

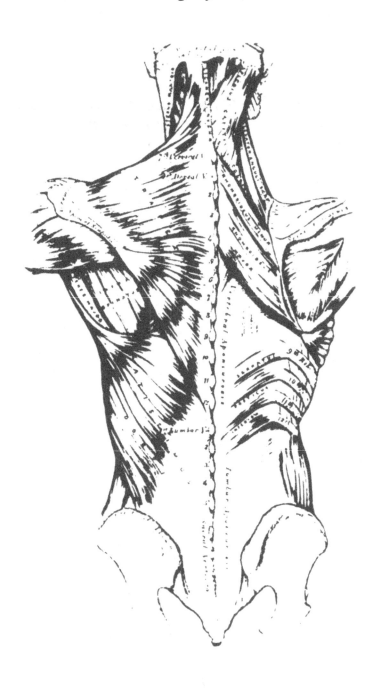

July

And July in her eyes hath place.
> —*Thomas Morley*
> *Madrigals to Four Voices (1594)*

July 1

d.1568

Levinus Lemnius, Dutch physician and chemist, student of *Vesalius*; he wrote one of the first descriptions of the distillation of brandy.

1966

Medicare inaugurated in the United States.

July 2

d.1566

Michel de Notre Dame (*Nostradamus*), French physician, astrologer, and alchemist. As a physician he distinguished himself by his fearless treatment of plague victims. As a prophet his rhymed quatrains have interested occultists for centuries.

1739

Jacob de Castro Sarmento receives his doctorate in medicine at Aberdeen. *De Castro* is one of the Jewish families that furnished a large number of distinguished physicians, many of whom suffered under the Inquisition.

d.1843

Samuel Hahnemann, German physician. His *Organon der rationellen Heilkunde* (1810) described his "like cures like" system of therapeutics, otherwise known as homeopathy.

117

July 3

Feast of *Raymond Lully* (1235–1315), Franciscan (third order) physician, alchemist, philosopher, lay theologian, reformer, missionary, and (possible) martyr. He was a prodigious alphabetizer and, in addition to his works on medicine and philosophy, wrote a manual on chivalry.

1569

William Chamberlen, a Huguenot physician, emigrates from France to England; he and his family would keep the invention of obstetrical forceps a family secret for over a century.

d.1920

William Gorgas, American military physician who acted on the theory that yellow fever was transmitted by mosquitos bred in standing water in the Panama Canal Zone, and thus largely defeated the disease.

July 4

d.1840

Karl Ferdinand von Gräfe, Polish surgeon and ophthalmologist, the Father of Modern Plastic Surgery. In 1818 he perfected the first satisfactory technique of rhinoplasty; he also contributed to the development of blood transfusion, cesarean section, and cleft palate repair. His son *Albrecht* was one of the greatest ophthalmalogic surgeons of the century.

d.1934

Madame Curie dies a victim of the radioactive substances she had discovered.

1954

The wife of *Dr. Samuel Sheppard*, an osteopathic neurosurgeon, is murdered; forensic serology findings that would have cleared him are ignored at the trial.

July 5

1650

Thomas Willis (1621–1675) writes the first description of manic-depressive psychosis.

1948

Britain's National Health Insurance goes into effect.

July 6

d.1784

Jean Baptiste Aymen, French physician who used turnips to prevent scurvy; he also wrote on Hippocratic medicine. *Linnaeus* named the plant species *Aymenea* in his honor.

1885

Nine-year-old *Joseph Meister* receives the first injection of weakened microbes of *Pasteur's* hydrophobia vaccine.

1921

Richard Pampuri (1897–1930)graduates at the top of his medical class. In 1927 he entered the Order of St. John of God, and in 1969 was proclaimed a saint.

Pasteur

July 7

d.1568

William Turner, English physician, botanist, naturalist, and priest; in 1544 he wrote the first modern ornithological work.

b.1768

Philip Syng Physick, the Father of American Surgery. His teacher, *John Hunter,* pointed out several cadavers to Philip's father with the comment, "These are the books your son will learn under my direction. The others are fit for very little." ***Dr. Physick*** did not like to write, but he invented the first tonsillotome, was among the first to use a stomach pump, described rectal diverticula, developed an operation for an artificial anus, and pioneered the use of absorbable animal tissue ligatures. "His most famous operation was the successful removal of several hundred stones from the bladder of *Chief Justice Marshall.*"

b.1844

Camillo Golgi, Italian cytologist who described the interconnecting cytoarchitecture of the nervous system. Several cell types and a silver stain are named for him. He also traced the relationship between the stages of the life cycle of the malarial parasite and tertian and quartan fevers. He received the 1906 Nobel Prize.

b.1852

John H. Watson, British physician and author, wounded at the battle of Maiwand while serving as a military surgeon during the Second Afghan War (1878–1879).

July 8

b.1776

Dominique Jean Larrey, French military surgeon who invented field hospitals and "flying ambulances." In 1802 he called attention to the contagiousness of trachoma and in 1812 published the first description of trench foot. He espoused using maggots to treat wound infections and a humeral amputation was named for him. He was wounded three times in battle.

Larrey's "flying ambulance."

d.1781

Jean Baseilhac (Frère Côme), French physician and surgeon. In 1729 he joined *L'Ordre des Feuillants* and in 1753 established a hospital for the poor. He introduced lateral lithotomy, lithotome cache, and used the haut appareil for the suprapubic extraction of bladder stones.

1800

Benjamin Waterhouse vaccinates his son *Daniel.*

d.1939

Havelock Ellis, English physician, essayist, sexologist, and editor. His most famous works are *The Dance of Life* and the seven volume *Studies in the Psychology of Sex.* "A man's destiny stands not in the future but in the past. That, rightly considered, is the most vital of all vital facts. Every child thus has a right to choose his own ancestors." He was Jerry in H.D.'s *Palimpsest.*

July 9

d.1677

Johann Scheffler. This Lutheran physician became a Catholic priest and under the name *Angelus Silesius* earned a reputation as a mystical poet.

> *God is my final end;*
> *Does He from me evolve,*
> *Then He grows out of me*
> *While I in him dissolve.*

July 10

d.1099

Ruy Diaz de Bivar, "*El Cid*," Spanish warrior who in 1067 founded the first leprosarium.

d.1767

Alexander Monro, Scottish physician who founded the first medical school in Britain and began a medical dynasty of *Monros* in Edinburgh.

1775

James Thacher is accepted by the medical examiners for a military appointment in the Continental Army. He had had no formal training except as an apprentice to a physician. His *Military Journal* (1823) was a remarkable and important historical document. He was also a botanist and wrote the first American work on hydrophobia.

b.1873

Theodore Simon, French physician who with former medical student *Alfred Binet* (1857–1911) devised the first objective measure of intelligence in children, to show "the benefits of instruction for defective children." (1905)

July 11

Feast of *Saint Benedict* (c. 480–550), patron saint of all diseases.

d.1710

Domenico Guglielmini, Italian physician and mathematician who researched hydraulics and originated the law of constant interfacial angles for salt crystals.

d.1754

Thomas Bowdler. Unable to stand the sight of blood and suffering, he abandoned his London medical practice to devote his time to revising and enlarging his sister's *Family Shakespeare,* a version deprived of all earthiness. His name has as a result given a word for puritanical revision to the language.

July 12

1540

Formal incorporation of the Barber's Company and the Surgeon's Guild by an act of Parliament.

b.1813

Claude Bernard, the father of experimental physiology.

b.1849

William Osler, Canadian physician and author whose greatest influence was as a teacher in the wards.

b.1856

August Désiré Waller, French physiologist who was the first to use electrodes to study current patterns in the heart.

July 13

d.1629

Caspar Bartholin, primus, Swedish physician and theologian, who wrote on anatomy, green jasper, pygmies, and unicorns.

d.1793

Jean Paul Marat is murdered in his bath by *Charlotte Corday. Dr. Marat* had had an aristocratic practice and had written on venereal disease.

Jean Paul Marat

Skeleton of the 'Irish giant' obtained illegally by John Hunter

July 14

b.1728

John Hunter, English anatomist and collector, who founded pathological anatomy in England.

1883

The first issue of the *Journal of the American Medical Association* is published.

July 15

1532

Rabelais dedicates his *Hippocratis ac Galeni Libri Aliquot* to bishop *Geoffrey d'Estissac.*

1577

"The 15, 16, and 17 of July sickened...above 300 persons, and with 12 days space died an hundred Scholars, besides many Citizens....The Physicians fled...to save themselves and theirs...some thought that this Oxford mortality...was devised by the Roman Catholics, who used the Art Magick in the design." *Anthony Wood, History and Antiquities of the University of Oxford* (1647)

1622

Charter granted to the Royal Society of London.

July 16

b.1877

Bela Schick, Viennese-born American pediatrician who developed a test for diphtheria. "Children are not simply micro-adults, but have their own specific problems."

b.1902

Alexander R. Luria, Russian neurosurgeon whose work with World War II head injuries laid the foundation for his pioneering advances in brain localization and description of the frontal lobe syndrome.

July 17

b.1818

Ignatz Philipp Semmelweiss, Hungarian obstetrician who wrote *The Cause, Concept and Prophylaxis of Puerperal Fever* (1861). His martyrdom is dramatically described in *Morton Thompson's The Cry and the Covenant.* "Whenever the benefactors of mankind are mentioned, the name of Semmelweiss should rank pre-eminent."

July 18

Feast of *Saint Camillus de Lillis;* with *St. John of God,* this Capuchin is the patron of hospitals, nurses, and the sick.

b.1884

William Gordon Lennox, American physician and educator, an originator of modern treatment of epilepsy, remembered in the *Lennox-Gastaut syndrome,* one of the more complex seizure patterns in children.

b.1888

Louis Hopewell Bauer, pioneer in aeronautical medicine, editor-in-chief of the *Journal of Aviation Medicine* for almost a quarter of a century.

b.1894

Armand James Quick, a hematologist who developed the prothombin test named after him.

1929

Edward D. Churchill performs the first successful pericardial resection in the United States at the Massachusetts General Hospital on an 18-year-old schoolgirl with constrictive pericarditis diagnosed by *Dr. Paul Dudley White.*

July 19

d.1573

Dr. John Caius, English physician, Greek scholar, and Renaissance humanist. In his 1552 *A Boke, or Counseill Against the Disease Commonly Called the Sweate, or Sweatyng Sickness* he attacked quackery; he founded the Cambridge college that bears his name.

1832

The British Medical Association is founded.

b.1865

Charles H. Mayo, Founder of the Mayo Clinic.

b.1896

Archibald Joseph Cronin, Scottish physician and novelist. His most famous works include *The Keys of the Kingdom*, the medical *The Citadel*, and the autobiographical *Adventures in Two Worlds*.

d.1901

Pierre Carl Edouard Potain, French cardiologist, inventor of the portable sphygmomanometer.

July 20

Feast of *Saint Margaret*, patron saint of childbirth and children. (She probably did not exist.)

b.1839

Julius Friedrich Cohnheim, German pathologist and physiologist. He proposed the theory that neoplasms originated from embryonal cell rests.

July 21

d.1637

Daniel Sennert, German physician who performed the first authenticated cesarian section on a living woman.

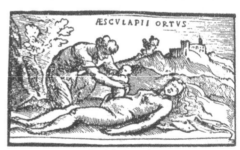

Cesarian section

1768

William Heberden communicates to the College of Physicians a paper entitled *Pectus Dolor or Angina Pectoris*. It was published in 1772.

b.1835

Robert Munro, Scottish physician and archeologist.

d.1901

Felix Joseph Henri de Lacaze-Duthiers, French anatomist, biologist, zoologist, and professor of malacology; the father of experimental zoology in France.

b.1913

Catherine Cole Storr, English psychiatrist, novelist, playwright, and author of *Clever Polly* stories. Under the pen name of *Irene Adler*, she wrote *Freud for the Jung* (1963).

July 22

b.1889

Morris Fishbein, American physician. In his *Doctors and Specialists*, *Dr. Fishbein* noted that *primum non nocere*, usually translated as "first, do no harm," could also be translated as "do not be naughty in the first place," or even as "no harm in just once."

d.1893

John Rae, physician and Arctic explorer.

1955

Male nurse *John Armstrong* murders his five-month-old son Terence with barbiturates (Seconal); this crime becomes the first sensational trial for murder by use of barbiturates.

July 23

d.1368

Guy de Chauliac, French surgeon, the Father of Surgery *(Fallopius)*. He wrote the earliest description of femoral hernia. "We are like children sitting on the neck of a giant who see all that he sees and something besides." His usage of this Bernardian theme places him tenth in the transmission of *Robert K. Merton's* Otsogian aphorism.

Guy de Chauliac

b.1773

Abraham Colles, Irish physician, anatomist, and surgeon. He wrote *Surgical Anatomy* (1811), had a fracture, a ligament, a fascia, and a law named for him, and performed the first ligature of the innominate artery. "Be assured, that no man can know his own profession perfectly, who knows nothing else."

b.1872

Edward Adrian Wilson, English physician and explorer who studied Antarctic penguins. He perished on the return from the South Pole.

July 24

d.1842

Dominique Jean Larrey, military surgeon, inventor of the ambulance volante.

d.1842

Pierre Joseph Pelletier, French pharmacologist. He and his long-time collaborator, *Joseph Bienaimé Caventou*, were known as "the Siamese twins of pharmacy." Together they isolated emetine, strychnine, carnine, brucine, veratrine, quinine, colchicine, and caffeine. They also gave chlorophyll its name.

July 25

1778

John Hunter publishes his *The Natural History of the Human Teeth* together with *A Practical Treatise on the Diseases of the Teeth*. *Hunter* invented the use of braces for the correction of dental malocclusion.

d.1816

Franz Kaspar Hesselbach, German surgeon and anatomist. He specialized in hernias and described *Hesselbach's triangle*—an area bounded by the inferior epigastric artery, the rectus abdominis, and the inguinal ligament.

d.1865

Dr. James Barry, retired army surgeon, at age 73 after a long and distinguished career in the West Indies, South Africa, and India. She had matriculated at age 16 (1808) at Edinburgh University and acquired the reputation of a ladies' man and duellist. Her secret was not discovered until after her death.

b.1978

Louise Brown, the first artificially conceived (outside of the womb) fetus. Her birthweight was 5 pounds, 12 ounces. The gynecologist was *Patrick Steptoe* and the physiologist was *Robert Edwards.*

b.1880

Joseph Moscati. This Neapolitan physician treated *Enrico Caruso* and is a canonized saint in the Catholic church.

July 26

b.1875

Carl Gustav Jung, Swiss psychiatrist, the founder of analytical psychology. He originated the concept of the collective unconsciousness, introduced the terms introvert, extrovert, anima, animus, and synchronicity, and was a student of the occult, yoga, witchcraft, and alchemy.

Saint Pantaleon

1895

Maria Sklodowska marries *Pierre Curie*.

b.1908

Salvador Allende Gossens, Chilean physician and politician. He became the first freely elected Marxist president in the Western hemisphere.

July 27

Feast of *Saint Pantaleon,* physician and martyr, patron saint of those afflicted with headaches.

1660

Niels Stensen matriculated at the University of Leiden.

b.1851

Graham Steell, English physician. The diastolic murmur of pulmonic insufficiency was described by him in 1888 and is named for him.

1893

Dr. Mary Corinna Putnam marries *Dr. Abraham Jacobi. Dr. Jacobi* was the father of American pediatrics and wrote a history of American pediatrics. *Dr. Putnam* was referred to by *William Osler* as a "modern Trotula, a woman in the profession with an intellect so commanding that she will take rank with the *Harveys,* the *Hunters,* the *Pasteurs,* the *Virchows,* and the *Listers.*" (*Madame Trotte de Salerne* was a legendary female medicus of the School of Salerno.) *Dr. Putnam* was the first woman admitted to the Ecole de Médecin at Paris (from which she graduated with honors) and the foundress of the Association for the Advancement of the Medical Education for Women.

d.1931

August Henri Forel, Swiss physician, psychologist, and entomologist. He studied brain anatomy and the social psychology of ants and described some 3500 new species of Hymenoptera.

July 28

1865

Dr. Edward Pritchard, a Glasgow practitioner, was hanged for the murder of his wife and mother-in-law. It was the last public execution in Scotland.

July 29

1782

Dr. Benjamin Rush advises physicians to care for the poor, go regularly to church, and take their fees in goods and produce. But he warns them to never resent an affront offered by a patient, never become intimate with a patient, never sue a patient, never dispute a bill, never make light of a patient's complaint, and never appear to be in a hurry.

1811

The laying of the cornerstone of New York's Bellevue Hospital.

b.1841

Armauer Gerhard Henrick Hansen, Norwegian physician who discovered the bacillus that causes leprosy (*Hansen's disease*).

b.1849

Max Simon Nordau, German physician and Zionist author, critic, and dramatist. His favorite topic was degeneracy.

July 30

d.1566

Guillaume Rondelet, French physician and naturalist, friend and coworker of *Rabelais.* His *Libri de piscibus marinus* includes a description of 250 fish species. He provided *Rabelais* the model for the antifeminist *Rondibilis* ("Roly-poly") in Book III.

1965

President Lyndon B. Johnson signs *Medicare,* a medical-hospital insurance plan for the aged, funded by Social Security.

July 31

b.1826

Nicholaus Friedreich, German physician who had a hereditary spinal ataxia and the disease of paramyoclonus multiplex named after him.

1865

Ferdinand Hebra tricks *Ignaz Semmelweiss* into a local madhouse where he dies within two weeks of the equivalent of puerperal sepsis.

August

August 1

d.1708

Edward Tyson, English physician and anatomist who introduced the concept of the "missing link" into anthropology; he also wrote the first important work on comparative morphology.

1822

William Beaumont begins his pioneering experiments and observations on the stomach of *Alexis St. Martin.*

August 2

b.1802

August Bérard, French anatomist and innovative surgeon; the suspensory ligament of the pericardium is named for him.

1869

The first successful removal of a diseased human kidney is performed by the German surgeon and gynecologist, **Gustav Simon** (1824–1876). He also reintroduced splenectomy.

d.1922

Alexander Graham Bell, inventor who in 1886 was given an honorary degree in medicine from the University of Heidelberg for his development of a telephone probe to locate foreign bodies; the development of X-rays rendered this instrument obsolete, and its most famous trial on the corpse of the assassinated **President Garfield** was a failure. **Bell** also invented a "vacuum jacket" to administer artificial respiration to premature babies (his two sons died of respiratory distress syndrome of prematurity); this was never used in such infants, but was the prototype for the "iron lung" later employed with polio victims. He also founded the leading American research journal, *Science.*

August 3

1859

The American Dental Association is founded.

August 4

1443

Hôtel-Dieu is chartered.

b.1794

Hans Karl Leopold Barkow, German anatomist. Several carpal and tarsal ligaments are known by his name.

Hôtel-Dieu

1906

The American pathologist *Howard Taylor Ricketts* (1871–1910) publishes his paper *"The transmission of Rocky Mountain spotted fever by the bite of the wood tick."* Rickettsia are named for him.

August 5

d.1880

Ferdinand von Hebra, Viennese dermatologist who described erythema multiforme, pityriasis rubra, and prurigo ferox, all of which are named for him.

d.1940

Frederick Albert Cook, American physician and explorer who claimed to have reached the North Pole on April 21, 1900. He also claimed the first ascent of *Mount McKinley (Denali)* in 1906, but his "summit pictures" were proved to have been taken elsewhere.

1968

Publication of "A Definition of Irreversible Coma: Report of the Ad Hoc Committee of the Harvard Medical School to Examine the Definition of Brain Death."

August 6

d.1553

Girolamo Fracastoro, Italian physician, biologist, musician, mathematician, geographer, astronomer, and poet. He studied medicine with *Copernicus,* was a professor of logic and a classical scholar. He was official physician to the Council of Trent, which moved to Bologna after an outbreak of plague. His poem *"Syphilis sive morbus gallicus"* gave the name to that disease, and his *De contagione et contagiosis morbis* (1546) was the first complete statement of the nature of contagion, disease germs, and the modes of transmission of infectious diseases.

Girolamo Fracastoro

b.1828

Andrew Taylor Still, the American founder of osteopathy. He started the American School of Osteopathy in Kirksville in 1892, the *Journal of Osteopathy* in 1894, and wrote *The Philosophy of Osteopathy* (1899) and *Osteopathy, Research and Practice* (1910).

b.1881

Alexander Fleming, Scottish bacteriologist who discovered lysozomes and penicillin.

August 7

1495

First mention of recognizable syphilis in Emperor Maximilian's Edict against Blasphemers.

1962

President *John F. Kennedy* presents a gold medal for distinguished civilian federal service to *Frances Oldham Kelsey*, the FDA physician who blocked the sale of thalidomide in the United States. "The Lady Who Said No" resisted rigorous demands to clear the drug.

August 8

Feast of *Saint Cyriacus*, child martyr, patron saint of those afflicted with mental diseases or possession.

1786

M. G. Paccard, a local physician, won the de Saussure prize for the first ascent of Mont Blanc.

d.1828

Karl Pieter Thunberg, Swedish physician, botanist, and naturalist. He succeeded *Linnaeus* as professor of botany; his name is the eponym for the tropical plant genus *Thunbergia.*

1846

The Geneva Convention is signed protecting wounded and medical personnel and medical supplies in wartime.

1966

At Methodist Hospital in Houston, Texas, *Dr. Michael DeBakey* installed the first successful artificial heart pump (a left ventricular bypass) in a patient.

August 9

b.1819

William Thomas Morton, the American dental surgeon who introduced ether as a general anesthetic.

1867

Joseph Lister formally introduces the antiseptic principle in surgery in an address to the British Medical Association.

August 10

Feast of *Saint Lawrence*, patron saint of those with disorders of the back.

b.1786

Jean Guillaume Auguste Lugol, French dermatologist who developed an iodine solution for the treatment of skin conditions.

August 11

b.1861

James Bryan Herrick, American physician who wrote the first descriptions of blood-cell sickling and of coronary thrombosis. His "Clinical Features of Sudden Obstruction of the Coronary Arteries" appeared in the *JAMA* in 1912. "The doctor may also learn more about the illness from the way the patient tells the story than from the story itself." *Memories of Eighty Years*

d.1961

Richard Lewisohn, surgeon who in 1915 discovered a blood preservative (sodium citrate) that made blood banks possible.

August 12

Feast of *Saint Claire*, patron saint of ophthalmia. *St. Claire's disease* is conjunctivitis or any disease of the eye.

d.1560

Thomas Phayre, English pediatrician. "A general scholar and well versed in the Common Law" he wrote a book *Of the Nature of Writs* and translated *Virgil;* "But the study of the Law did not fare well with him which caused him to change his Copy and proceed Doctor in Physic." *(Fuller, Worthies of England)* "Thus endeth ye boke of childerne composed by *Thomas Phayer* Studiouse in Philosophie and Physike." (1545)

b.1762

Christoph Wilhelm Hufeland, German eclectic physician, journal editor, and pragmatic debunker of fads; he argued strongly against the temporary insanity plea. "If the physician presumes to take into consideration in his work whether a life has value or not, the consequences are boundless and the physician becomes the most dangerous man in the state."

d.1810

Etienne Louis Geoffroy, French physician, otologist, and entomologist who classified the order Coleoptera.

1865

The first successful use of carbolic acid antisepsis on an 11-year-old boy with a compound fracture of the left leg at the Glasgow Royal Infirmary. In his 1867 *Lancet* article *Joseph Lister* notes that he was first struck by the effect of carbolic acid on the sewage of Carlisle in 1864.

Joseph Lister

August 13

b.1625

Erasmus Bartholin, Danish physician and physicist who discovered the double refraction of Icelandic feldspar (calcite).

d.1826

René Laennec, French army doctor and physician, who invented the stethoscope.

d.1865

Ignaz Phillip Semmelweis, heroic advocate of antisepsis as the answer to the scourge of puerperal fever, and who himself died of an infection from a gynecological lesion.

August 14

d.1552

Fra Paolo Sarpi, Venetian statesman and historian, Servite monk and friend of **Galileo, Harvey,** and **Bacon.** He wrote many scientific treatises; his anatomical discovery was the contractility of the iris. He used the newly invented telescope to draw the first map of the moon. His classic *History of the Council of Trent* (1619) showed him first a patriot and a religious reformer second; his loyalty was to Venice above the Papacy.

1767

King's College Medical School, New York, is founded.

d.1819

Erik Acharius, Swedish physician and botanist who studied cryptograms and wrote a fundamental work on lichens.

1874

The German physician, *Adolf Kussmaul* (1822–1902), publishes his paper on coma and air hunger in diabetic acetonuria. In his study of aphasia he coined the term "word blindness."

1991

Howard B. Dean, physician and lieutenant governor of Vermont, succeeds to that state's highest office upon the unexpected death of Governor Richard Snelling.

August 15

d.1610

Peter Lowe, "Scotchman, Arellian, Doctor in the Facultie of Chirurgerie in Paris, and Chirurgion Ordinaire to the Most Victorious and Christian King of Fraunce and Navarre." He was a Marianist and a pirate. The 2nd edition of his book, *The Whole Course of Chirurgerie,* included the first English translation of **Hippocrates.** In 1599 he founded the Royal College of Physicians and Surgeons of Glasgow, the only royal college to unite medicine and surgery under one governing authority.

b.1840

Richard von Krafft-Ebing, German neuropsychiatrist who wrote the classic *Psychopathia Sexualis.*

1951

The Provençal village of Pont Saint Esprit in southern France goes mad with ergotism, *St. Anthony's fire (ignis sacer or mal des ardents)*— a syndrome of delirium, hallucinations, nausea, colic, vomiting, insomnia, and a burning sensation (in the anus). Fungus-infected rye bread contained an LSD-like hallucinogen that affected 300 people, 50 of whom went insane and three died.

August 16

Feast of *Saint Roche; St. Roche's disease* was bubonic plague, cholera, or the primary bubo of syphilis.

b.1626

Peter Chamberlen, inventor of the first practical obstetrical forceps, a family secret for a century.

b.1816

Johann Jacob Guggenbühl, a pioneer in the residential treatment of cretinism at Abendberg; he extolled the curative effects of the beauty of Nature.

Saint Roche

b.1832

Wilhelm Max Wundt, German physician and psychologist who became the founder of experimental psychology.

1898

The German surgeon *Dr. August Bier* administers the first spinal anesthetic.

August 17

b.1878

Oliver St. John Gogarty, Irish physician, poet, conversationalist, and politician; he was the model for *Buck Mulligan* in *James Joyce's Ulysses*—a fact that broke up his long-time friendship with *Joyce*. "The telescope, the microscope and the test-tube have made skeptics of us all. We have changed wisdom for an exact knowledge of stains, precipitants, reactions and refractions, and put it, for this generation at least, beyond recall." *I Follow Saint Patrick*

1938

The first successful ligation of a patent ductus arteriosus is performed in a 7-year-old girl by *Dr. Robert Gross* at Boston Children's Hospital.

August 18

1321

Ordinance to burn all lepers in France; they were found guilty of treating with the Saracens to poison Christians everywhere.

1840

The American Society of Dental Surgeons is founded.

August 19

1603

Dated dedication of *Thomas Lodge's Treatise of the Plague*. The book was printed at the author's expense and distributed to the needy; in 1625 *Dr. Lodge* died of the plague.

August 20

1897

Ronald Ross (1857–1932) identifies the Anopheles mosquito as the insect vector of malaria.

d.1915

Paul Ehrlich, German bacteriologist. *Ehrlichiosis* is an extremely rare infectious disease.

d.1915

Carlos Juan Finlay de Barres, Cuban physician who founded the doctrine of mosquito-borne diseases, especially yellow fever.

August 21

1784

Benjamin Franklin mentions bifocals in a letter to *George Whately.*

b.1826

Karl Gegenbaur, German anatomist. *Gegenbaur's cells* are osteoblasts.

b.1867

Edward Ernest Maxey, American otolaryngologist who in 1899 gave the first description of Rocky Mountain spotted fever.

August 22

1485

First rumors of the English sweating sickness (*sudor Anglicus*).

d.1643

Johann Georg Wirsung, Bavarian anatomist who was assassinated in Padua in a duel with a Belgian. The pancreatic duct bears his name.

d.1828

Franz Joseph Gall, German neuropsychologist who founded phrenology, the belief that mental abilities relate to the shape of the skull. He distinguished gray from white matter in the brain and demonstrated the decussation of the pyramids; with his student *J.C. Spurzheim,* he traveled to America. His skull is preserved.

August 23

b.1768

Astley Paston Cooper, English surgeon who in 1801 performed the first myringotomy to relieve deafness. His name is recalled in a number of anatomical eponyms.

1927

The anarchists *Sacco* and *Vanzetti* are executed. Their doom was sealed by forensic ballistics performed by *Dr. Calvin Goddard,* former cardiologist and deputy director of the Johns Hopkins Hospital.

August 24

1940

The Australians *Howard Florey* and *Ernest Chain* publish "Penicillin as a chemotherapeutic agent" in the *Lancet.*

August 25

b.1793

Martin Heinrich Rathke, Polish pathologist, physiologist, zoologist, and anatomist; *Rathke's pouch* is a depression in the roof of the embryonic mouth in front of the bucco-pharyngeal membrane.

August 26

1766

Doctor of Medicine is conferred on *Johann Peter Frank* (1745–1821), a founder of public hygiene and the medical police. His six volume treatise, *System einer vollständigen medizinischen Polizei* (1779–1817), is a classic.

b.1842

Heinrich Irenaeus Quincke, German physician who discovered the lumbar puncture; *Quincke's sign* uses the capillary and venous pulses in the diagnosis of aortic insufficiency, and *Quincke's disease* is angioneurotic edema.

1857

William Witherspoon Ireland (1832–1909) is reported killed before Delhi. Actually he was wounded by a bullet that destroyed one eye and passed around the base of the skull to the ear on the opposite side. He was the medical superintendent of the Scottish National Institution for Imbecile Children at Lambert from 1869 to 1879. His *Idiocy and Imbecility* (1877) was the first organized medical textbook on mental deficiency; he published essays in other fields, including history, literature, psychology, philosophy, and languages.

d.1910

William James, American psychologist. He was the model for Dr. Austin Sloper in his brother Henry's novel, *Washington Square*.

August 27

d.1574

Bartolomeo Eustachi, Italian physician and anatomist; several eponyms remember his name.

1761

Oliver Goldsmith concludes his review of *Leopold Auenbrugger's* treatise on percussion with the quotation, "If it cannot cure, it can do no harm."

1872

Dr. Robert Battey performs his first bilateral oopherectomy. *Battey's operation* was a treatment for dysmenorrhea, amenorrhea, epilepsy, mania, and nymphomania; it was attacked as "spaying."

1918

The first case of Spanish influenza occurs in Boston; 25% of the American population became ill and a half million died in the worst pandemic to afflict humankind since the Black Death—it killed 1% of the world's population.

August 28

1844

The American physician *Dr. Washington L. Atlee* (1808–1878) performs the first successful myomectomy.

b.1828

William Alexander Hammond, the court-marshalled Surgeon General of the United States, who wrote the first description of athetosis.

b.1878

George Hoyt Whipple, American pathologist who won the 1934 Nobel Prize for showing that pernicious anemia could be controlled with a diet of liver.

August 29

b.1632

John Locke, English philosopher and physician. In 1677 *Locke* wrote the first description of trigeminal neuralgia, and joined *Sydenham* in urging the use of cinchona. ("Jesuit's bark" was disliked by Protestants because of its name.) "Against *Locke's* philosophy I think it an unanswerable objection that, although he carried his throat about with him in this world for seventy-two years, no man ever condescended to cut it." *de Quincy,* "On Murder Considered as One of the Fine Arts"

b.1809

Oliver Wendell Holmes, American physician, anatomist, and writer. "If all medicine were very costly, and the expense of it always came out of the physician's fee..." "How could such a people be content with any but 'heroic' practice?" "A medicine...should always be presumed to be hurtful....I firmly believe that if the whole materia medica, as now used, could be sunk to the bottom of the sea, it would be all the better for mankind— and all the worse for the fishes."

Oliver Wendell Holmes

b.1911

John Charnley, the major innovator of the artificial joint.

August 30

Feast of *Saint Fiacre,* patron saint of those who suffer from piles and hemorrhoids, and of gardeners and horticulturists. He was also a misogynist and so became the patron of those afflicted with venereal disease. The Paris cab gets its name from a church with his name on it.

b.1705

David Hartley, English physician, psychologist, and philosopher. He was one of the originators of the doctrine of associationism and physiological psychology.

d.1804

Thomas Percival, English physician who introduced cod liver oil into therapeutics in 1782. He wrote on lead poisoning and smallpox. His *Medical Ethics* remains a classic.

August 31

b.1821

Hermann Helmholtz, a German physicist, anatomist, and physiologist. His *Description of an Eye Mirror for the Examination of the Retina in the Living Eye* (1850) documents his discovery of the ophthalmoscope; he was the first person to see inside a live human eye. He discovered the first law of thermodynamics, introduced the graphic method for studying muscle contraction, and measured the speed of nerve impulse conduction. *Helmholtz's ligament* is the axial ligament of the malleus of the ear.

Hermann Helmholtz

b.1841

Edward Gamaliel Janeway, American physician who inaugurated systematic autopsies at Bellevue Hospital.

b.1870

Maria Montessori, the first woman in Italy to take the degree of Doctor of Medicine; she devised a whole new system of education for the young based on principles that *Itard* and *Seguin* had used with the retarded.

1909

Paul Ehrlich injects compound 606, "the magic bullet," into a syphilitic rabbit. This would become the drug arsphenamine or Salvarsan.

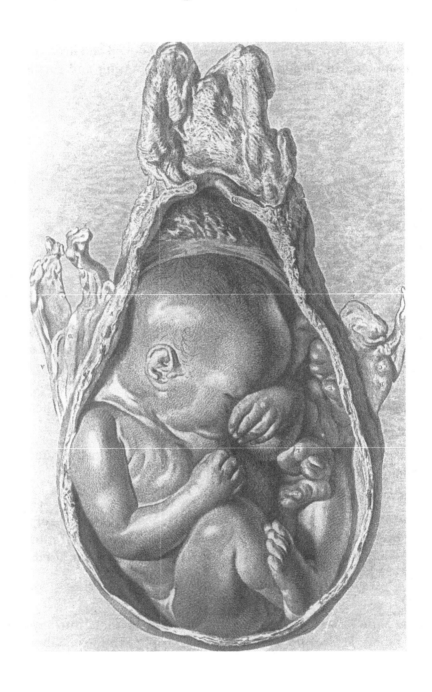

September

What tune the enchantress plays
In the aftermaths of soft September.

— *A. E. Housman, Last Poems*

September 1

Feast of **Saint Giles**. *St. Giles' disease* is cancer or leprosy.

b.1832

Ephraim Cutter, American physician, inventor of the laryngoscope.

1849

Outbreak of the Broad Street pump cholera epidemic that would be investigated by *John Snow*.

1858

Anatomy, Descriptive and Surgical is published by **Henry Gray** (1825–1861).

September 2

b.1730

John Cochran was a British surgeon's mate during the French and Indian Wars and the last medical director of the Continental Army.

b.1800

Willard Parker, American surgeon. He was the first American to operate successfully on an abscessed appendix.

d.1917

Revere Osler, son of **William Osler**, great-grandson of **Paul Revere**, mortally wounded at Ypres.

1952

First cardiac surgery in which the deep-freezing technique is used: *Dr. Floyd John Lewis* operates on a five-year-old girl at the University of Minnesota at Minneapolis.

September 3

b.1643

Lorenzo Bellini, Italian physician and anatomist, and professor of philosophy and anatomy at Pisa. His *De structura renum* (1662) described a number of kidney structures that bear his name. His *Gustus organum* (1665) studied the organs of taste and his *De urinis et pulsibus* (1683) related pulse and fever.

b.1860

William Barnes, American surgeon and entomologist who owned the largest collection of Lepidoptera then in existence (several hundred thousand species).

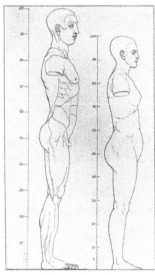

*Figures showing
male and female proportions*

September 4

b.1791

Robert Knox, Scottish anatomist and ethnologist. His professional career was ruined in an 1828 scandal:

> *Burke's the butcher, Hare's the thief,*
> *Knox the man that buys the beef.*

In 1852 he published two beautiful monographs, *Manual of Artistic Anatomy* and *Great Artists and Great Anatomists.*

d.1940

Hans Zinsser, American physician, microbiologist, philosopher, and poet. He isolated the germ causing typhus and developed a vaccine. He wrote *Rats, Lice and History,* and his auto-biography, *As I Remember Him,* closes with a sonnet:

> *Now is death merciful. He calls me hence*
> *Gently, with friendly soothing of my fears...*

d.1965

Albert Schweitzer, musician, philosopher, and medical missionary.

September 5

b.1633

Bernardino Ramazzini of Carpi, Italian physician who wrote the first systematic and comprehensive study of occupational diseases, *De morbis artificium* (1700); it included descriptions of lead and antimony poisoning.

September 6

b.1714

Robert Whytt, Scottish physician who wrote on neurological subjects. *Whytt's reflex* is the abolition of the pupillary contraction to light.

1791

Quaker physician and philanthropist *John Coakley Lettsom* writes in a letter that medicine "is not a lucrative profession. It is a divine one." He founded the Medical Society of London, wrote a *Natural History of the Tea-Tree* and a *History of the Origin of Medicine;* he kept a journal of *Captain Cook's* second voyage. He pioneered in the study of drug abuse and alcoholism and wrote the first description of alcoholic multiple neuritis. The *Lettsomian Lectures* are named in his honor. Aware of the social context of disease, he once wrote to the local government about a destitute patient, "Money, not physic, will cure her." He was also blessed with a sense of humor:

September

When people's ill, they comes to I,
I physics, bleeds, and sweats 'em;
Sometimes they live, sometimes they die.
What's that to I? I let's 'em.

(Epigram on himself)

b.1876

John James Richard Macleod, Canadian physiologist, joint discoverer of the insulin treatment for diabetes.

1891

H.C. Dalton performs the first suture of the pericardium during an operation on a 22-year-old man stabbed in the chest.

September 7

b.1677

Stephen Hales, vicar of Teddington. He was the first to measure blood flow, blood volume, and blood pressure in his *Statical Essays* (1731–1733).

b.1816

Ferdinand von Hebra, pioneer dermatologist.

1888

Edith Eleanor McLean is born weighing 2 pounds, 7 oz, in New York City. She is the first premature infant to be put in an Auvard incubator or "hatching cradle."

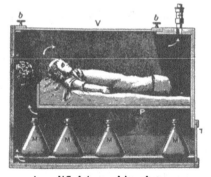

A modified Auvard incubator

d.1922

William Stewart Halsted, American surgeon, one of the founders of the Johns Hopkins Medical School. The first pair of rubber surgical gloves were made at his order for his chief operating room nurse, **Caroline Hampton**. Otherwise her eczema would have forced her to retire; she later became his wife. His early research into the use of cocaine as a local anesthetic led to his addiction; he broke this habit by becoming a lifelong morphine addict, with no apparent impairment to a very active career. His example would seem, however, to be an exception not to be followed. Witness the self-deception in the following passage:

> "I, the unfortunate *Doctor Polyakov*, who have become addicted to morphine in February of this year, warn anyone who may suffer the same fate not to attempt to replace morphine with cocaine. Cocaine is a most foul and insidious poison....It does not in the least affect my capacity to work...and I give my professional oath that my addiction causes no harm to my patients." **Mikhail Bulgakov, A Country Doctor's Notebook**

d.1936

Berkeley George Andrew Moynihan, British abdominal surgeon who popularized duodenal ulcers and introduced the concept "hunger pain" felt by patients with such ulcers several hours after eating. "Statistics...will prove anything, even the truth."

September 8

d.1637

Robert Fludd, English physician, philosopher, and inventor. He was a Paracelsist and a Rosicrucian, one of the first vaccinators and pulse timers, and the inventor of an automatic lyre.

d.1734

Michel Sarrazin, French-Canadian physician and botanist. His is the eponym for the carniverous Pitcher plant—*Sarracenia purpurea*.

1867

Dr. Samuel A. Mudd, a Confederate sympathizer who had set the leg of *John Wilkes Booth* and had been sentenced to life imprisonment as a result, is released from his cell at Fort Jefferson, Florida, to help with an outbreak of yellow fever. He was pardoned by President *Andrew Johnson* a year and a half later.

1935

ENT specialist *Dr. Carl Austin Weiss* assassinates the "Kingfish," Senator *Huey P. Long* of Louisiana, and is killed by *Long's* bodyguards.

September 9

d.1771

Tobias Smollett, English physician and novelist. He was surgeon's mate on an expedition against Spanish sea power in the Americas. He wrote *Roderick Random* and *Peregrine Pickle,* translated Cervantes, and was nicknamed Smelfungus by *Laurence Sterne.*

1896

Ludwig Rehn of Frankfort City Hospital performs the first successful suture of a human heart wound in the right ventricle of 22-year-old stabbing victim, *Wilhelm Justus.*

1904

James Joyce and *Oliver St. John Gogarty* move into the Martello tower in Sandycove, the setting for the first scene of *Ulysses.*

September 10

From *Benjamin Rush's* notebook: "Thank God! Out of one hundred patients whom I have visited or prescribed for this day, I have lost none."

1624

Thomas Sydenham baptized. "There were cases where I have consulted my patient's safety and my own reputation by doing nothing."

b.1771

Mungo Park, Scottish physician and explorer. He pioneered in the exploration of the Niger River and discovered eight new species of fish in Sumatra. He was one of the models for Gideon in *Sir Walter Scott's Chronicles of the Canongate* (1827). "There is no creature in Scotland that works harder and is more poorly requited than the country doctor, unless perhaps it may be his horse."

b.1775

John Kidd, English physician and naturalist. He discovered naphthalene in coal tar, leading to the use of coal as a source of chemicals.

b.1842

Henry Granger Piffard, American physician, author of the first systematic treatise on dermatology in America.

1876

Frantisek Chvostek, Sr., Austrian surgeon, reports on a new sign of unilateral facial irritability and spasm on tapping in postoperative tetany.

September 11

d.1851

Sylvester Graham, Connecticut clergyman, physician, and eccentric nutritional reformer. *Emerson* satirized this dietician as the "poet of bread and pumpkins"; he gave his name to a cracker. Grahamism was an almost religious movement.

1892

Cholera panic on Long Island, New York; an armed mob prevents foreigners from landing on Fire Island.

1952

An artificial aortic valve designed by *Dr. Charles Anthony Hufnagel* is fitted to a patient at Georgetown University Medical Center.

September 12

d.1681

John Ward, English parson who studied medicine. His *Diaries* contain many valuable references to the medicine, surgery, and science of his day.

1878

The oldest public monument in London, the Alexandrian obelisk "Cleopatra's Needle," is erected through the philanthropy of England's first dermatologist, *Erasmus Wilson* (1809–1884).

1916

First experimental induction of pellagra by diet restriction in convict volunteers is performed by *Joseph Goldberger* and *G.A. Wheeler.*

b.1917

Suyin Han, Chinese physician and novelist, author of *A Many-Splendoured Thing.* "Truth, like surgery, may hurt, but it cures."

September 13

d.1321

Dante Alighieri, Italian poet and member of the apothecary's guild.

1813

Richard Bright graduates from medical school. In his *Reports of Medical Cases* (1827) he will give a classic description of nephritis, or Bright's Disease.

Dante Alighieri

b.1851

Walter Reed, American physician, head of the United States Army Yellow Fever Board.

September 14

b.1486

Heinrich Cornelius Agrippa, German physician and occultist. His *On the Vanity of the Sciences* condemned the medical profession for its part in the delusion of witchcraft and cruelty to the insane.

b.1849

Ivan Pavlov, Russian physician, physiologist, behavioral psychologist, and Nobelist.

1902

Dr. Luther Hill of Montgomery, Alabama, becomes the first American to successfully suture a heart wound; his patient was a 13-year-old boy.

September 15

d.1613

English essayist and poet *Sr. Thomas Overbury* is murdered because of the sentiments expressed in his poem, *A Wife*. His becomes the first recorded trial of a medical man for a serious crime in England. *Lord Rochester* and the *Countess of Essex* who ordered the murder by poison (rosalgar) are pardoned; the doctors involved, *Richard Weston* and *James Franklin*, are executed.

1884

The use of cocaine as a local anesthetic is first demonstrated at the Heidelberg Congress of Ophthalmology.

d.1875

Guillaume B.A. Duchenne, French neurologist. A type of muscular dystrophy and a progressive bulbar paralysis are named after him. He invented the muscle biopsy in 1868; *Gowers* called it an "histological harpoon." In 1862 he compiled an atlas of photographs depicting facial expression mechanisms in different states of emotion. *Charcot* said, "How is it that, one fine morning *Duchenne* discovered a disease that probably existed in the time of *Hippocrates?*...-Why do we perceive things so late, so poorly, with such difficulty?"

b.1921

Gordon Ostlere, English surgeon and anesthesiologist. Under the pen name of *Richard Gordon* he wrote the humorous *Doctor in the House* series.

September 16

b.1651

Englebert Kaempfer, German physician, botanist, and collector. He was "the first German explorer, the *Humboldt* of the 17th century, the scientific discoverer of Japan." He wrote a *History of Japan*.

b.1796

Jean Baptiste Bouillaud, French cardiologist and neurologist. He was the last great bloodletter and the first to localize the speech center to the middle of the left cerebral hemisphere in his *Traité...de l'encéphalite* (1825).

d.1819

John Jeffries, American physician and balloonist. He delivered the first public lecture on anatomy in New England in 1789 and completed the first air crossing of the English Channel in 1785.

d.1916

Thomas Lauder Brunton, English physician and pharmacologist.

d.1936

Jean-Baptiste Charcot, French physician and explorer. He did research in ethnography, geophysics, hydrography, meteorology, microbiology, bacteriology, terrestrial magnetism, glaciology, nautical studies, and mapping the Antarctic.

September 17

Feast of *Saint Hildegard of Bingen*, a Benedictine Abbess of the Convent of Rupertsberg near Bingen. She was a physician, physiologist, scientist, preacher, theologian, painter, poet, composer, dramatist, and visionary. Among several medical works she wrote, the *Liber Compositae Medicinae* is most famous. She set one of her morality plays to music and so has some claim to having written the first opera.

1530

François Rabelais matriculates at the University of Montpellier School of Medicine.

b.1883

William Carlos Williams, New Jersey pediatrician and poet, author of *Paterson* (5 volumes, 1946–1958) and *In the American Grain* (1925), and other key 20th century works. "It's the humdrum, day-in, day-out, everyday work that is the real satisfaction of the practice of medicine...the actual calling on people, at all times and under all conditions, the coming to grips with the intimate conditions of their lives...has always absorbed me. I lost myself in the very properties of their minds: for the moment at least I actually became *them*....For the moment I myself did not exist, nothing of myself affected me." *Autobiography* (1948).

d.1991

Frank H. Netter, American medical illustrator.

September 18

1888
The American Pediatric Society is founded.

September 19

1782
Harvard Medical School is founded.

b.1934
Richard Raskind, American ophthalmologist and tennis coach to *Martina Navratilova. Dr. Raskind* is unique in having lost US championships in both the men's and women's divisions, the latter under the name *Renee Richards*. Her autobiography was titled *Second Serve.*

September 20

d.1576
Girolamo Cardano (Jerome Cardan), Milanese physician, mathematician, and astrologer. He identified typhus under the name *morbus pulicaris*, wrote a major mathematical treatise on algebra, *Ars Magna*, but was always a bit of a madman. "In the diagnosing of diseases *don Girolamo* has not an equal!" "I deemed medicine a profession of sincerer character than law."

September 21

1507
Sudor Anglicus, the sweating sickness, breaks out in London.

1948
Phillip Showalter Hench and *Edward C. Kendall* administer compound E (cortisone) to patients with arthritis.

September 22

d.1506

Marco Antonio Della Torre, Italian anatomist who employed *Leonardo da Vinci* as his illustrator. He and his physician father are commemorated in the bronze bas-reliefs of the Church of San Fermo in Verona, sculpted by *Andrea Riccio.* In pleading before the Bishop of Padua to mitigate the punishment of a learned monk who had gotten five nuns pregnant, this physician responded to the cleric's question, "What answer shall I give to God on the Judgment Day when he says to me, Give an account of thy stewardship?" with the clever retort, "My lord, say what the Evangelist says: Lord, thou deliveredst unto me five talents: behold I have gained beside them five talents more." (*Castiglione, The Book of the Courtier*).

d.1957

Oliver St. John Gogarty, Irish physician, politician, and writer.

September 23

1518

The Royal College of Physicians is chartered to conduct qualifying examinations.

d.1657

Joachim Jung, German physician and naturalist who wrote on geometry and logic.

d.1738

Hermann Boerhaave, physician and teacher, the "Dutch Hippocrates." His *Institutiones medicae in usus annuae exercitationis domesticos digestae* (Leyden, 1708) emphasized the chemical factors in digestion. He was an eclectic systematist, as shown by his *Aphorisms* (1709); he was an accomplished musician and the founder of the

Botanical Garden at the University of Leyden. His pupil *Haller* called him *communis totius Europeae praeceptor.* "My best patients are the poor because the Lord has taken it upon Himself to pay me for them."

d.1939

Sigmund Freud, Austrian neurologist and psychiatrist. "After forty-one years of medical activity, my self-knowledge tells me that I have never really been a doctor in the proper sense. I became a doctor through being compelled to deviate from my original purpose; and the triumph of my life lies in my having, after a long and round-about journey, found my way back to my earliest path. I have no knowledge of having had in my early years any craving to help suffering humanity....I scarcely think, however, that my lack of genuine medical temperament has done much damage to my patients."

September 24

1810

In a long letter to the Managers of the Pennsylvania Hospital, *Benjamin Rush* outlines an advanced and insightful approach to the institutional management of mental patients. He later wrote *Medical Inquiries and Observations upon the Diseases of the Mind* (1812), the first American psychiatry textbook.

b.1501

Gironimo Cardano Cardano, Italian physician and mathematician, author of The Book of the Game of Chance. "My Nativity: Although various abortive medicines were tried in vain...."

d.1541

Aureolus Phillippus Theophrastus Bombastus ab Honenheim (Paracelsus). 'There are two kinds of physicians—those who work for love, and those who work for their own profit. They are both known by their works."

> *"Can all your Paracelsian mixtures cure it?"*
>
> *(Middleton, Faire Quarrell)*

September 25

1493

Christopher Columbus leaves Cadiz for his second voyage to the New World. *Diego Alvarez Chanca,* physician to *Ferdinand* and *Isabella,* is appointed fleet surgeon; his letter to the city of Seville will become the major record of that exploration. He made the selection for the site of the first permanent settlement in Haiti.

b.1613

Claude Perrault, French physician, anatomist, architect (the east facade of the Louvre), and translator of *Vitruvius.* His brother *Charles* wrote perdurable fairy tales.

d.1773

Agostino Bassi, Italian physician and civil servant. He wrote on pellagra, cholera, contagion, vinification, cheese, and the cultivation of potatoes. In 1835 he proved that muscardine (mal del segno or calcino, a disease of silkworms) was caused by a fungus, now named *Botrytis bassiana* in his honor. He thus anticipated *Pasteur* by a decade in proposing that disease can be caused by animal or plant parasites.

d.1890

Jesse William Lazear, American physician, student of *Osler;* he died helping to show that mosquitoes carry yellow fever.

September 26

Feast of *Saints Cosmas* and *Damien,* twins and physicians who practiced without asking fees.

Sts. Cosmas and Damian

b.1668

Giorgio Baglivi, Italian physician who first distinguished smooth from striated muscle. As a prominent iatrophysicist, he divided the body machine into many smaller machines. Nevertheless, his *De praxi medica* (1696) showed him to be an eminently practical clinician, observer, and teacher.

September 27

b.1849

Ivan Petrovich Pavlov, Russian physician and behavioral physiologist. He developed a "theory of types" to describe personality differences among dog subjects and utilized the ancient Hippocratic terms—Sanguine, Choleric, Phlegmatic, and Melancholic. This physiological classification of genetically determined types of learning behavior was thoroughly expurgated from official Soviet Psychology.

b.1887

Alan Brown, Canadian pediatrician, co-inventor of Pablum, the royalties from which helped to build the Hospital for Sick Children in Toronto into a major center.

September 28

b.1789

Richard Bright, English physician, botanist, geologist, naturalist, artist, and traveler.

b.1809

Alvan Wentworth Chapman, American physician and botanist. During the Civil War he helped Union soldiers escape from prison. He wrote *Flora of the Southern States* (1860), and the genus *Chapmania* is named in his honor.

d.1895

Louis Pasteur, French chemist and bacteriologist; "the most perfect man that has ever entered into the kingdom of science." *(Osler)*

d.1946

Theodore Lasater Terry, American ophthalmologist. *Terry's disease* is retrolental fibroplasia.

September 29

1686

"And this is about the sum of all I know respecting the cure of diseases, up to the day on which I write—namely, the 29th September, 1686." *Thomas Sydenham, Schedula Monitoria de Novae Febris Ingressa. Sydenham* popularized the use of cinchona bark for the treatment of malaria and introduced the terms scarlet fever, scarlatina, and whooping cough.

b.1850

Étienne-Louis Arthur Fallot, French physician who described maladie bleue or morbus caeruleus in 1888. This became known as the *tetralogy of Fallot* despite the fact that it had been previously described by *Peacock* in 1858.

b.1903

John Gibbon, developer of the heart-lung machine used in open-heart surgery.

d.1927

Willem Einthoven, pioneer in the development of the electrocardiograph.

September 30

1846

T.G. Morton painlessly extracts a tooth from *Eben H. Frost* under ether anesthesia.

d.1894

Vladimir Betz, Russian anatomist who in 1874 described the giant pyramidal cells of the motor cortex named for him.

d.1943

Richard Austin Freeman, English surgeon and health official (at the Holloway prison). He created *Dr. John Evelyn Thorndyke* and based this fictional detective on *Alfred Swaine Taylor,* a medical jurist and authority on poisons. "Thorndyke is a unique figure in the legal world. He is a barrister and a doctor of medicine. In the one capacity he is probably the greatest criminal lawyer of our time. In the other he is, among other things, the leading authority on poisons and on crimes connected with them; and so far as I know, he has never made a mistake." *As a Thief in the Night*

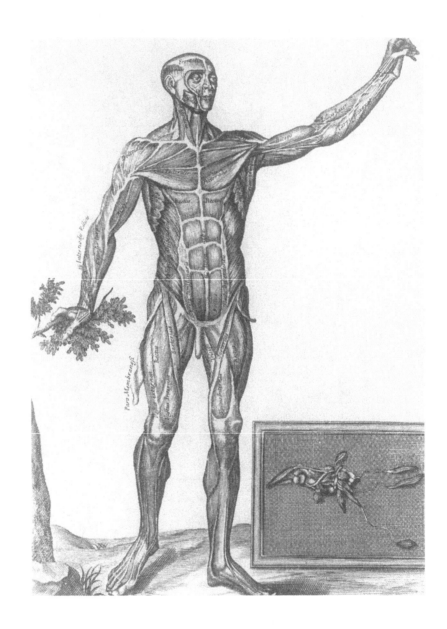

October

Then summer fades and passes, and October comes.

—*Thomas Wolfe*
You Can't Go Home Again

October 1

1853

John Langdon Haydon Langdon-Down (1828–1896) becomes a medical student at The London Hospital. Noted for his gentleness, his 1866 "Observations on an ethnic classification of idiots" gave the first detailed description of the genetic disorder that bears his name. He also described dystrophia adiposa-genitalis four decades before *Fröhlich*.

d.1863

Ebenezer Emmons, American physician and geologist who pioneered in Paleozoic stratigraphy.

1915

Six of *Joseph Goldberger's* prison volunteers placed on restrictive diets develop pellagrous lesions.

October 2

1861

William J. Little (1810–1894) presented his classic paper on asphyxia and brain damage (cerebral palsy, Little's disease) before the Obstetrical Society of London. The society president thought teething a more likely cause of paraplegia. Paper titles took the place of abstracts: "On the influence of abnormal parturition, difficult labours, premature birth, and asphyxia neonatorum, on the mental and physical condition of the child, especially in relation to deformities."

1893

Opening date of classes for the first students at Johns Hopkins Medical School.

October 3

b.1777

James Jackson, American physician, therapeutic nihilist. In 1822 he published the first description of peripheral alcoholic neuritis. "I have often remarked that, though a physician is sometimes blamed very unjustly, it is quite as common for him to get more credit than he is fairly entitled to; so that he has not, on the whole, any right to complain." *Letters to a Young Physician* (1855)

b.1854

William Crawford Gorgas, American surgeon. As chief sanitation officer for the Panama Canal, he directed yellow fever research there.

1878

Thomas Morgan Rotch (1849–1914), Boston pediatrician, reports a new sign—the absence of resonance in the fifth right intercostal space in pericardial effusion. *Rotch* introduced a method of percentage feeding of fat, sugar, and protein for infants.

October 4

Feast of *Saint Francis of Assisi*. *St. Francis' Fire* is erysipelas.

b.1716

James Lind, Scottish physician, the father of naval hygiene in England.

1774

Benjamin Church (1734–1777), the first Surgeon General of the Continental Army, is found guilty of treason. In addition to being a physician, he was a high-living minor poet and adulterer; the decoding of an intercepted cryptogram proved his undoing.

b.1883

William Gibson Arlington Bonwill, American physician, surgeon, dentist, inventor, carpenter, cabinet maker, and school teacher. His inventions include the all-porcelain tooth crown, the dental and surgical engine, the removable bridge, a safety-pointed pin, and a machine to carve marble and rock.

d.1989

Graham Chapman, physician, cofounder of *Monty Python's Flying Circus.*

October 5

1823

The first issue of the *Lancet.*

d.1885

Daniel A. Carrión, the patron saint of Peruvian science; the Daniel Carrión School of Medicine in Lima is named for him. As a sixth year medical student, he died proving the identity of Peruvian warts (*verruga Peruanna*) with Oroya fever. They are now known as *Carrión's disease.* The classic paper, "Concurso sobre Veruga peruana," appeared in *Monit. méd.,* Lima, four days earlier.

1978

The *New England Journal of Medicine* declares professional courtesy obsolete.

October 6

b.1510

John Caius, English physician who appeared as a character in *Shakespeare's Merry Wives of Windsor.* "*Master Caius,* that calls himself doctor of physic...*Master Doctor Caius,* the renowned French physician."

b.1777

Guillaume Dupuytren, French physician and surgeon. Kidnapped in childhood, a self-made millionaire, baron of the Empire, royalist, "the First of surgeons and the least of men." He invented an early form of stomach tube and had a fracture, a contracture, and several anatomical structures named for him.

d.1799

William Withering, English physician, botanist, and mineralogist. He introduced the use of digitalis (foxglove) into medicine. *Witheringia,* a genus of Solanaceae, is named in his honor.

d.1855

Elisha Bartlett, American physician who wrote a classic book on *Fevers* (1842). "Long abused humanity is likely at no very remote period to be finally delivered from the abominable atrocities of wholesale and indiscriminate drugging."

1947

Bobby Brown ends the Dodger lead in the last game of the World Series. He was in four World Series with the New York Yankees, later went to medical school, and eventually became president of the American League.

François Magendie

October 7

d.1885

François Magendie, French physician physiologist who discovered the *Bell-Magendie Law* involving the anatomical and functional discrimination of sensory and motor neurons. He introduced iodine, bromine, strychnine, and morphine compounds into general practice.

d.1939

Harvey Cushing, American neuro-surgeon, neuroendocrinologist, and medical historian. He died of a heart attack suffered while lifting a heavy copy of *Vesalius* in his library.

October 8

Harvey Cushing

Feast of *Saint Triduana of Scotland.* A local prince desired her because of her beautiful eyes; she removed them and sent them to him. She is the patroness of all those who suffer diseases of the eyes.

d.1729

Richard Blackmore, English physician and notoriously bad poet, "the father of Bathos, and indeed the Homer of it"; "How few have reached the low sublime." *(Swift)* In his *A Treatise upon the Small Pox* (1723) *Blackmore* related an anecdote of *Thomas Sydenham* the interpretation of which is still debated today: "When one Day I asked him to advise me what Books I should read to qualify me for Practice, he replied, Read *Don Quixot,* it is a very good Book, I read it still. So low an Opinion had this celebrated Man of Learning collected out of the Authors, his Predecessors."

b.1888

Ernest Kretschmer, German psychiatrist who related body build and constitution to character and mentality.

October 9

d.1562

Gabrielo Fallopius, Italian anatomist, surgeon, and botanist. He was especially interested in reproductive anatomy and introduced the terms ovarian tube, vagina, clitoris, placenta, cricoid, labyrinth, and ciliary body. He wrote *Observationes anatomicae* (1561) and was the first to use an aural speculum to diagnose ear disease.

b.1860

Leonard Wood, American physician, military governor of Cuba, Congressional Medal of Honor winner for the campaign against *Geronimo.*

d.1967

Ernesto "Che" Guevara, physician and revolutionary. "Each and every one of us will pay, on demand, his part of sacrifice."

October 10

d.1425

Ugolino di Montecatini, Italian physician, one of the earliest balneologists.

1673

Appointment of *Dr. Hugh Chamberlen* as Physician-in-Ordinary to His Majesty.

1789

Dr. Joseph I. Guillotin (1738–1814) proposes to the French National Assembly the use of a new method of capital punishment without torture. It will be named after him—and later used on him.

d.1884

Robert Christian Barthold Ave-Lallement, German physician who explored Brazil and the Nile.

d.1945

Walter B. Cannon, American physician who pioneered the concept of homeostasis and the application of X-rays to the diagnosis of gastrointestinal diseases. His *Bodily Changes in Pain, Hunger, Fear, and Rage* (1915) and *The Wisdom of the Body* (1932) are classics, and his autobiographical study of serendipity, *The Way of an Investigator,* was corrected on his deathbed.

October 11

b.1593

Nicolaes Tulp, Dutch physician and anatomist. He wrote the first Amsterdam pharmacopia and *Observationes Medicae* (1641). He described beriberi, the lacteals, and the ileocecal valve (*Tulp's valve*), and dissected a chimpanzee. He is the central figure depicted in **Rembrandt's** *The Anatomy Lesson.*

Nicolaes Tulp

1670

William Lilly, English uroscopist and astrologer is licensed Doctor of Medicine. He was satirized as "Sidrophel" in **Butler's Hudibras.**

October 12

1884

Epitaph in Morristown, New Jersey, cemetery:

> *In memory of*
> **Charles H. Salmon,**
> *who was born September 10th, 1858.*
> *He grew, waxed strong, and developed into*
> *a noble son and loving brother. He came to*
> *his death on the 12 of October, 1884, by the*
> *hand of a careless drug clerk and two excited*
> *doctors, at 12 o'clock at night in Kansas City.*

b.1879

René Lériche, French surgeon, father of modern vascular surgery. He introduced periarterial sympathectomy (*Lériche's operation*) and described *Lériche's syndrome* —thrombotic occlusion of the distal abdominal aorta with symptoms of pain or coldness in the legs, intermittent claudication, and impotence. In his *Souvenirs de ma vie morte* (1956) he relates a visit to the Hunter Museum in which *Sir Berkeley Moynihan* showed him an anatomical fragment that he diagnosed as "a piece of pitted intestine with a dysenteric perforation." The specimen was from *Napoleon Ier.*

d.1915

English nurse *Edith Cavell* is executed as a spy for helping Allied soldiers escape into occupied Belgium.

October 13

1805

Hanaoka Seishu (1760–1835), Japanese physician, performs an operation for breast cancer using an anesthetic, tsusensan, 37 years before *Long* and 42 years before *Simpson.* The research that led to this innovation is depicted in the novel *The Doctor's Wife* by *Sawako Ariyoshi.*

b.1821

Rudolph Virchow, German physician, politician, and archeologist. His *Cellular Pathology* (1858) was a seminal work. He was known as "der kleine Doktor" because of his stature. "Medicine is a social science in its very bone marrow."

b.1881

Russell Lafayette Cecil, American physician pioneer in rheumatic diseases, founder and president of the *American Rheumatism Association,* and editor of a classic textbook of medicine.

October 14

d.1667

Francis Glisson, English physician, anatomist, and physiologist. His *De rachitide* (1650) was one of the earliest descriptions of rickets, and his *Anatomia hepatis* (1654) included a description of the liver capsule that bears his name. He was "the most accurate anatomist that ever lived."

October 15

d.1564

Andreas Vesalius of Brussels, ship-wrecked on Zante.

1783

Jean Pilâtre de Rozier, French physician–apothecary, makes the first human ascent in a balloon. On June 15, 1785, he will become the first balloon fatality.

Andreas Vesalius

1923

Paul Drucker, Copenhagen pediatrician, invents the heelstick.

October 16

b.1708

Albrecht von Haller, Swiss physiologist with special interests in the vascular and nervous systems. "*Albrecht the Great*," "a master physiologist," he was a child prodigy, a wunderkind, a student of *Boerhaave*, a botanist, anatomist (the *tripus Halleri* is the branching of the celiac axis), bibliographer, medical historian, and lyric nature poet—his *Versuch schweizerischer Gedichte* was published anonymously in 1732.

1846

A twenty-year-old printer, *Gilbert Abbot*, is administered ether by *William Morton* and then operated on by *Dr. John C. Warren* at the Massachusetts General Hospital. *Dr. Warren*: "Gentlemen, this is no humbug." *Dr. Henry Bigelow:* "I have seen something to-day that will go around the world."

October 17

b.1886

Ernest W. Goodpasture, American immunologist who developed and did not patent the technique of live chick egg embryos to cultivate vaccines. He isolated the mumps virus and during the 1918 influenza pandemic and described the association of pulmonary hemorrhage with glomerulonephritis *(Goodpasture's syndrome)*.

d.1887

Robert Hunt, English physician and physicist who pioneered in photography and film development.

October 18

Feast of *St. Luke*, "dear and glorious physician," author of two books of the *New Testament.*

St. Luke

1645

Daniel Whistler submits his thesis for his medical degree at the University of Leyden: *De Morbo puerili Anglorum quem patrio idiomate indigenae vocant the Rickets.*

d.1842

The three-year-old daughter of American physician, poet, and dramatist, *Thomas Holley Chivers* (1809–1858) dies. The memorial poem he composed "To Allegra Florence in Heaven," was later plagiarized by *Edgar Allen Poe* for his "The Raven." *Poe* described *Chivers* as "one of the best and one of the worst poets in America."

1854

The British war cabinet unanimously confirms *Florence Nightingale* as Superintendent of the Female Nursing Establishment of the English General Hospitals in Turkey.

October

Benjamin Jowett, the Oxford Platonist, on being asked what his lady-love, **Florence Nightingale**, was like, replied, "Violent, very violent."

d.1871

Charles Babbage, English mathematician who constructed the first simple ophthalmoscope. He devised the basic principles of modern computers, computed the first reliable actuarial tables, designed the modern postal system, and constructed the first speedometer.

October 19

b.1605–d.1682

Sir Thomas Browne, English physician and writer, author of *Religio Medici* (1643), and *Pseudodoxia epidemica (Vulgar Errors)* (1646), and *Hydriotaphia (Urn Burial)* (1658). "For the World I count it not an Inn but a Hospital, a place not to live, but to dye in." (1682)

1944

Dr. Clarence Crafoord, chief surgeon of Sabbatsberg Hospital in Stockholm, performs the first successful surgical correction of coarctation of the aorta.

Sir Thomas Browne

October 20

d.1524

Thomas Linacre, "the type of the literary physician" *(Osler)*. He founded the College of Physicians in 1518 and in 1520 left the prac-

179

tice of medicine to become a Catholic priest. As a classical scholar, *Thomas More* and *Desiderus Erasmus* were among his pupils. "He was one of the first English Men that brought polite learning into our Nation, and it hath been justly questioned by some of the Goliaths of learning, whether he was a better Latinist or Grecian, or a better Grammarian or Physician." (*Anthony à Wood*)

b.1616

Thomas Bartholin, Danish physician and scholar. He gave the first description of the thoracic duct and the lymphatic system and in 1672 wrote one of the first scientific treatises on diseases in the Bible. (The Swede *Olof Rudbeck* independently discovered the lymphatic system.)

1801

"On the Education of a Man in the Wild, or the First Physical and Moral Developments of the Young Sauvage de l'Aveyron" by *Dr. Jean-Marc-Gaspard Itard* was published.

b.1812

Austin Flint, American internist who pioneered in the use of the binaural stethoscope and described the presystolic murmur of aortic regurgitation that is named for him. (1862) *"The American Laennec"*

1867

Eduardo Bassini receives a groin wound while fighting with Garibaldi's troops. His search for treatment leads this Italian surgeon to lay the foundation for modern herniorrhaphy.

October 21

d.1558

Julius Caesar Scaliger, Italian physician and philosopher, scholar, scientist, and polymath.

1639

The earliest Virginia law having specific reference to the medical profession is passed: "An act to compel physicians and surgeons to declare on oath the value of their medicines."

October 22

d.1578

Isabeau Rolant, wife of *Jehan Bony* of the Monceaux near St. Gervais, where the sign of the Red Rose hangs, dies of a tumor of eight years' duration. The autopsy report by **Ambrosie Paré** in *On Tumors in General* describes a hydatidiform mole.

ANNO·ÆTATI
68

Ambroise Paré

d.1889

Philippe Ricord, Baltimore-born French physician whose penchant for extravagant practical jokes cost him his internship with *Dupuytren.* He corrected *John Hunter's* erroneous identification of syphilis and gonorrhea. *Oliver Wendell Holmes* described him as "the *Voltaire* of pelvic literature—a skeptic as to the morality of the race in general, who would have submitted Diana to treatment with his mineral specifics and ordered a course of blue pills for the vestal virgins."

1910

In the Old Bailey a guilty verdict is passed on *Dr. Hawley Harvey Crippen* for the murder of his wife Cora. The wireless telegraph was used for the first time in the dramatic apprehension of the culprit at sea; his mistress accompanied him disguised as a young man. This case represented one *Dr. Bernard Spilsbury's* dramatic successes for forensic pathology.

October 23

1814

Joseph Constantine Carpue (1764–1846) performs the first plastic surgery operation in Britain. He reconstructs the nose of an army officer who lost it through mercury poisoning.

b.1837

Moritz Kaposi, Hungarian dermatologist who described *Karposi's sarcoma* (multiple bluish nodules in the skin marked by hemorrhages) and *Kaposi's disease* (brown spots and ulcers on the skin). Until the advent of AIDS Kaposi's sarcoma was one of the rarest of skin disorders.

b.1844

Robert Bridges, English physician, poet laureate. He wrote *The Testament of Beauty* (1929) and developed his own phonetic spelling.

b.1942

Michael Crichton, American physician and novelist; he wrote *The Andromeda Strain* and *The Terminal Man.*

1988

Pope John Paul beatifies *Niels Stensen,* Danish anatomist.

October 24

b.1632

Antony van Leeuwenhoek, pioneer Delft microscopist.

b.1909

John Milton McLean, American ophthalmologist who established the first corneal eye bank.

Antony van Leeuwenhoek

October 25

d.1653

Theophraste Renandot, French physician and editor, the "Father of French Journalism." He founded, edited, and published the first political journal, *La Gazette de France,* created the first free medical clinics for the poor, and started the first pawnshops in France.

b.1693

Antoine Ferrein, French surgeon and anatomist. He described *Ferrein's pyramids* in the kidney cortex and bile canaliculi, and originated the term vocal cord by comparing the laryngeal ligaments to violin strings.

October 26

1831

First appearance of Asiatic cholera in England at Sunderland.

1877

Joseph Lister repairs and wires a fractured kneecap using antisepsis in a procedure requiring a simple fracture to be converted to a compound one.

October 27

d.1553

Michael Servetus Villanovanus, Spanish anti-Arabist physician, and theologian. His *Christianisimi restitutio* both attacked the Trinity and described for the first time the pulmonary circulation. The Calvinists burned him at the stake for heresy in Geneva (on the issue of the Trinity). "I will burn, but this is a mere event. We shall continue our discussion in eternity" was his reply to his judges.

b.1761

Matthew Baillie, Scottish pathologist, nephew of the Hunters. His *The Morbid Anatomy of Some of the Most Important Parts of the Human*

Body (1793) was the first publication on pathology as a separate subject. He is credited with the first descriptions of cirrhosis of the liver, transposition of viscera, and gastric ulcer.

b.1900

Peter James Kerley, Irish radiologist who described *Kerley B lines* in pulmonary edema.

October 28

Feast of *Saint Jude,* patron saint of hopeless cases.

1631

Guy Patin writes to Falconet: "The apothecaries are in a mad rage about *La Médecin charitable* and its adherents who have medicines prepared cheaply at home." The book referred to was written by the physician *Philibert Guybert* in 1623.

b.1858

Henry Koplik, American pediatrician who isolated the bacillus for whooping cough and established the first milk depot for poor infants. *Koplik's spots* are an early diagnostic sign of measles.

d.1880

Henry Koplik

Edouard Seguin, French American physician who did ground-breaking work in mental retardation and clinical thermometry. He was the first president of the American Association on Mental Deficiency and one of the leaders of the institutional movement in the United States.

1880

Nature prints a letter from the Scottish physician *Henry Faulds* working in Tokyo on the utility of fingerprints in criminal investigation. "The pattern was unique."

1918

"The Plague of the Spanish Lady," an influenza epidemic, reached its height in Britain.

October 29

d.1692

Melchisédech Thévenot, French physician, Oriental linguist, librarian. He invented a new way to measure altitude.

d.1937

Theodore Simon Flatau, German otolaryngologist who studied infant vocalizations, voice failure in singers, and ventriloquism.

October 30

d.1850

Takano Choei, Japanese physician, Dutch scholar, translator, tippler, writer, traitor for advocating greater commerce with the West. He was thought dead until he published a translation the quality of which was his signature. He committed suicide when cornered by the police.

b.1895

Gerhard Domagk, father of sulfonamide therapy.

d.1910

Jean Henri Dunant, founder of the International Red Cross.

d.1956

Pio Baroja, Spanish physician and novelist. Basque, banned trilogist, and sometime baker.

October 31

1749

The English chemist *John Dalton* reads a paper with the first description of color blindness *(Daltonism)*.

b.1795

John Keats, English poet. He was apprenticed to a surgeon in 1810, registered as a medical student at Guy's Hospital in 1815 and qualified as licentiate of the Society of Apothecaries in 1816. "I am sure I could not take fees—and yet I should like to do so; it's not worse than writing poems and hanging them up to be fly-blown on the Review shambles."

d.1832

Antonio Scarpa, Italian anatomist and surgeon, chief surgeon to Napoleon, Father of Italian ophthalmology. He wrote the first full description of club foot (1803) and his eponyms include the sliding hernia, the ganglion of the vestibular nerve, the femoral triangle and the abdominal fascia. He was a stern professor at Padua:

> *Scarpa is dead;*
> *And I should care.*
> *He lived like a hog*
> *And died like a dog.*

b.1857

Axel Munthe, Swedish physician and writer. His bestselling book of reminiscences *The Story of San Michele* (1929) opens with a bargain with the devil—a secluded villa for the renunciation of ambition: "Will you at least leave me pity. I cannot live without pity if I am to become a doctor." "Yes, I will leave you pity, but you would have fared much better without it." The closing chapter is a surrealistic scene of final judgment in which the doctor brings his faithful dog to the gates of heaven, is defended by *St. Rocco*, the patron saint of dogs, and is finally saved by the intercession of *St. Francis* and a flock of birds he had protected from cruelty.

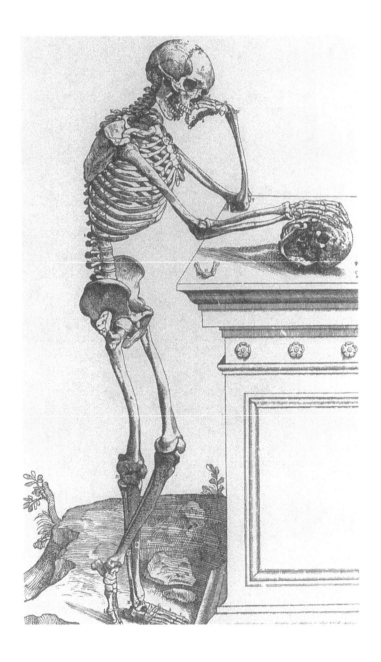

November's sky is chill and drear
November's leaf is red and sear
—Walter Scott, Marmion

November 1

1530

François Rabelais takes his bachelor's degree in medicine before the Faculty in Montpellier.

1532

François Rabelais is appointed physician to the Grand Hôtel-Dieu de Notre Dame de Pitié du Pont-du-Rhône in Lyon.

d.1714

John Radcliffe, American physician and philanthropist. To a shoemaker's wife who brought in her husband's urine specimen, he replied that if the shoemaker would fit him with a pair of shoes from a sample of his (Radcliffe's) urine, then he would give her a diagnosis. On his deathbed he said that when he first started practicing medicine, he had 20 remedies for each disease; when he finished, he had 20 diseases for which there was not a single remedy. *Osler* moralized, "One lesson learned from his life is that if you do not write, make money, and, after you finish, leave it to the Johns Hopkins Trust."

1848

The first medical school for women, the Boston Female Medical School, opens with an enrollment of 12. It was promoted by *Samuel Gregory*, a pioneer in medical education for women. It was the only college to graduate professionals with the title "Doctress."

At the Boston Female Medical School

1924

Henry Pancoast describes a syndrome of pain, *Horner's syndrome,* bone destruction and atrophy of hand muscles associated with a pulmonary sulcus tumor.

November 2

1322

Guilty verdict against *Jacoba Felice* (*Jacqueline Félicie de Almania*) who was charged by the masters in medicine at Paris of practicing medicine and surgery, of inspecting urine, feeling pulses, and touching limbs. In her defense she had argued that she entered into no contracts; patients paid what they wished when they got well. She treated mostly women and patients on whom the established physicians had already given up. She visited her patients frequently and enjoined them, "I shall make you well, God willing, if you will have faith in me." She was threatened with fine and excommunication if she persisted.

1664

First mention of plague in London's bills of mortality at the start of the Great Plague.

November 3

Feast of *Saint Zachary,* patron saint of those afflicted with mutism.

Feast of *Saint Hubert*, bishop, patron saint of hunters. *St. Hubert's disease* is hydrophobia.

Costume used by plague physicians

November 4

1846

Patent for the first artificial leg.

1862

Patent for the gatling gun awarded to **Richard J. Gatling** (1818–1903), North Carolina physician–inventor.

b.1862

Ernest Guglielminetti, Swiss physician who originated a method of tarring road surfaces and studied the physiological effects of altitude in balloons.

b.1865

Chevalier Jackson, American physician who perfected the use of the esophagoscope and the bronchoscope. He was a pioneer in peroral endoscopy. "All that wheezes is not asthma."

November 5

1650

Guy Patin is elected Dean of Faculty of Medicine at Paris. **Patin** (1601–1672) was an archconservative: "This powder of cinchona has not any credit here. Fools run after it because it is sold at a very high price, but having proved ineffective it is mocked now."

d.1714

Bernardino Ramazzini, Italian physician who recognized the importance of environmental influences on disease.

1858

Thomas Clifford Allbutt (1836–1925) enters the medical school of St. George's Hospital. **"Allbutt** of Leeds" invented the modern clinical thermometer (1866), incriminated arteriosclerosis as a cause of disease, and wrote on medical history in addition to his *System of Medicine* (8 volumes, 1896–1899). He was Tertius Lydgate in George Eliot's *Middlemarch*.

1892

William Forsyth Milroy, an Omaha physician, describes chronic hereditary edema (*Milroy's disease*).

d.1981

Samuel Rosen, American surgeon who first mobilized the stapes to treat otosclerosis—surgery on the smallest bone in the human body.

November 6

d.1852

Daniel Drake, peripatetic American pioneer physician, medical educator, poet, founder of two Ohio medical colleges, author of the humorous reminiscence *Pioneer Life in Kentucky.* He wrote the magisterial *A Systematic Treatise, Historical, Etiological and Practical, on the Principal Diseases of the Interior Valley of North America, as They Appear in the Caucasian, African, Indian and Eskimo Varieties of its Population,* a mixture of medicine, topography, meteorology, and ethnography. He was *e sylvis nuncius.*

b.1861

Dr. James Naismith, Canadian physician who invented basketball.

1880

French army surgeon *Charles Alphonse Laveran* (1845–1922) first observes the malarial parasite in human blood.

November 7

d.1599

Gasparo Tagliacozzi (*Taliacotius*), Italian plastic surgeon who wrote the first treatise on rhinoplasty, *De Curtorum Chirurgia. Tagliacozzi's* body was removed from its grave in the chapel of the nuns of San Giovanni Battista and reburied in unhallowed ground because the nuns claimed to have heard a voice above the grave proclaiming *Tagliacozzi's* damnation. The reason for ecclesiastical opposition to rhinoplasty may be deduced from the December 6, 1710, number of *The Tatler.* After an epigraph from Martial, *Non cuicunque datum est habere Nasum* (Not everybody can have a nose), **Steele** quotes *Hudibras:*

> *So learned Talicotius from*
> *The brawny part of porter's bum*
> *Cut supplemental noses, which*
> *Lasted as long as parent breech;*
> *But when the date of nock was out,*
> *Off dropp'd the sympathetic snout.*

and implies that the major reason for the loss of a nose was the "pox," thus making *Taliacotius* the first clap doctor—or perhaps the first love doctor.

1847

Elizabeth Blackwell's (1821–1910) name is entered on the roll books of the medical department of Geneva University.

b.1867

Marie Sklodovska, codiscoverer with her husband *Pierre Curie* of radium. They refused to take out patents on the usage of a substance that promised so much benefit to mankind.

Marie Curie

November 8

1847

James Young Simpson first uses chloroform on an obstetrical patient, *Wilhelmina Carstairs,* daughter of a physician.

1895

Wilhelm Konrad Roentgen X-rays his own hand.

b.1922

Christiaan Barnard, South African heart transplant surgeon.

November 9

b.1606

Hermann Conring (Conringius), German physician, naturalist, chemist, lawyer, and historian. *Conringia orientalis*, hare's ear mustard, is named in his honor.

b.1721

Mark Akenside, English physician and lyric poet who was satirized in *Smollett's Adventures of Peregrine Pickle*. He took one month to write his medical dissertation, *De Ortu et incremento foetus humani* (1744). He was arrogant, pedantic, and harsh of manner toward his poorer patients, a generally unsympathetic character. His politics were characterized by *Dr. Johnson* as an "impetuous eagerness to subvert and confound, with very little care what shall be established."

November 10

b.1493

Paracelsus, Swiss physician and alchemist. He was the first to describe the cardinal signs of inflammation: Rubor, Tumor, Calor, and Dolor—redness, swelling, heat, and pain. He espoused a "minerall physicke" or chemical therapy. His concept of the body's *archeus,* or ruler who pushes the body (back) toward a state of health, seems to be an anticipation of homeostasis.

d.1683

Robert Morison, British physician and botanist who provided the eponymous source for the genus *Morisonia*.

b.1728

Oliver Goldsmith, Irish physician, poet, playwright and novelist. After a youth of squalid dissipation, he learned to disregard accuracy and facts. His medical education was suspect.

> Goldsmith: *I do not practise. I make it a rule to prescribe only for my friends.*
>
> Beauclerk: *Pray, dear Doctor, alter your rule; and prescribe only for your enemies.*

This quondam "class clown" and village blockhead wrote one of the great novels in English literature (*The Vicar of Wakefield*), one of its great poems ("The Deserted Village"), and one of its great plays (*She Stoops to Conquer*).

> *When lovely woman stoops to Folly,*
> *And finds too late that men betray,*
> *What charm can sooth her melancholy,*
> *What art can wash her guilt away?*

b.1759

Friedrich Schiller, German physician and writer.

1871

Henry M. Stanley finds *Dr. David Livingstone* at Ujiji near Unyanyembe. *Livingston,* a medical missionary, had discovered Victoria Falls (1855), had explored the sources of the Nile, had described the relapsing fever that follows on the bite of a tick, and related the bite of the tsetse fly to a disease in cattle. Stanley became Alec MacKenzie in W. Somerst Maugham's *The Explorer.*

b.1882

Charles McMoran Wilson Moran, English physician who was criticized for including too much clinical detail in his 1966 book *Winston Churchill: The Struggle for Survival 1940–1965.*

November 11

Feast of *Saint Martin*. Delirium tremens was *St. Martin's evil.*

b.1742

Benjamin Shattuck, American physician and patriot. At the age of 60 he shouldered his musket and marched to Concord on April 19, 1775.

1762

The Pennsylvania Gazette announces a course of anatomic lectures by *William Shippen, Jr.*, a student of the *Hunters* and *Fothergill.* Later Surgeon General, he was persecuted for his anatomic studies.

b.1771

Ephraim McDowell, American frontier surgeon who performed the first ovariotomy.

b.1777

Marie François Xavier Bichat, a French histologist who founded tissue pathology. Although he described 21 tissue types, he disliked the use of the microscope. His career was meteoric; many anatomic structures are named for this short-lived genius. He published *Recherches sur la Vie et sur la Mort* in 1800 and *Anatomie générale* in 1802, the year he died.

d.1938

"Typhoid Mary" Mallon.

November 12

b.1793

Johann F. von Eschscholtz, German naturalist, anatomist, and explorer. The California poppy—*Echscholtzia californica*—is named for him, as is a bay in Kotzebue Sound, Alaska.

1846

Letter Patent No. 4848 issued to **Charles T. Jackson** (physician) and **William T.G. Morton** (dentist) for 10% of all profits on the use of ether in surgical operations.

d.1964

Henry Souttar, the first British surgeon to operate successfully within the heart in a 1925 operation on mitral stenosis. He also invented the *Souttar tube.*

November 13

b.1715

Dorothea Christiane Erxleben, German physician. She was the first woman medical doctor and practiced in Quedlinburg.

b.1866

Abraham Flexner, American medical educational reformer. His 1910 Carnegie Foundation report *Medical Education in the United States and Canada* criticized diploma-mill medical schools. (He was the brother of the American pathologist, **Simon Flexner.**)

November 14

1745

The first published account of lead poisoning, by Philadelphia surgeon **Thomas Cadwalader,** *An Essay on the West India dry-gripes . . . to which is added an extraordinary case in physick* (Philadelphia, *B. Franklin,* 1745). It was **Benjamin Franklin** who proved that the lead colic and palsy resulted from Jamaica rum being distilled through lead pipes.

b.1891

Frederick G. Banting, discoverer of insulin.

1901

Karl Landsteiner publishes "On Agglutination Phenomena of Normal Human Blood" in *Wiener klinische Wochenschrift.*

November 15

b.1313

Ibn Al-Khatib, Spanish physician.

1845

Florence Nightingale writes from Scutari "We have now *four miles* of beds, and not eighteen inches apart." She was a pioneer in the uses of social statistics and their graphical representation; she invented polar area charts ("coxcombs").

November 16

d.1785

Johann Gottschalk Wallerius, Swedish physician and mineralogist who did chemical research on fertilizers.

d.1958

Samuel Hopkins Adams, American writer who debunked quack nostrums in his *The Great American Fraud* (1906). "With the exception of lawyers, there is no profession which considers itself above the law so widely as the medical profession."

November 17

b.1645

Nicolas Lemery, French physician and chemist who discovered iron in the blood.

d.1863

Madame de Pauw dies of foxglove poisoning administered by *Dr. Couty de la Pommerais*. The toxicological solution to this murder by *Dr. Ambroise Tardieu* represented a milestone in forensic medicine.

November 18

b.1836

Cesare Lombroso, Italian physician and criminologist. His *L'uomo delinquente* (1876) and other works proposed a theory of "criminal types," a relationship between personality, physique, and mental constitution. He also carried out research on pellagra.

d.1887

Gustav Theodor Fechner, German physician, philosopher, psychologist, and anthropologist. He was the founder of experimental psychology, discovered the *Weber–Fechner law*, and identified *Mises' marginal plexus* of nerves in the eyelid (*Mises* was his pseudonym).

November 19

b.1722

Leopold Auenbrugger, Austrian physician who discovered percussion.

November 20

1911

Harvey Cushing writes to *Dr. Henry Christian*: "Why not put the surgical age of retirement for the attending surgeon at sixty and the physician at sixty-three or sixty-five, as you think best? I have an idea that the surgeon's fingers are apt to get a little stiff and this makes him less competent before the physician's cerebral vessels do. However, as I told you, I would like to see the days when somebody would be appointed surgeon somewhere who had no hands, for the operative part is the least part of the work."

1939

Howard W. Florey applies to the Rockefeller Foundation for a grant to study microbial antagonisms with substances including penicillin (approved and funded).

November 21

d.1555

Georg Bauer (Georgius Agricola), German physician and mineralogist, author of the classic *De re metallica.* He recognized industrial diseases and reintroduced ventilation into mines.

d.1775

"Sir" John Hill, English apothecary and botanist who also wrote plays, novels, satires, and a gossip column. He wrote the first book of British flora utilizing the new Linnaean nomenclature, the 26 folio volume *Vegetable System* with 1600 copper plate engravings of 26,000 different plants.

> *For physic and farces,*
> *His equal there scarce is,*
> *His farces are physic,*
> *His physic a farce is.*
> > *(David Garrick)*
>
> *The writer on snuff, valerian and sage,*
> *The greatest imposter and quack of his age;*
> *The punishment ordered for all such sad crimes,*
> *Was to take his own physic, and read his own rhymes!*

b.1785

William Beaumont, American frontier physiologist.

1846

Oliver Wendell Holmes writes to *Dr. Morton*: "My Dear Sir: Every body wants to have a hand in a great discovery. All I will do is to give you a list or two as to names—or the name—to be applied to the state produced and the agent. The state should, I think, be called 'Anaesthesia' . . . The adjective will be 'Anaesthetic.'...I would have a name pretty soon and consult some accomplished scholar...before fixing upon the terms, which *will be repeated by the tongues of every civilized race of mankind."* **Morton** settled on "Letheon."

November 22

1902

Walter Reed verifies the mosquito transmission of yellow fever.

November 23

b.1805

Joseph Pancoast, American surgeon who invented a new suture for plastic surgery and devised *Pancoast's operation,* the division of the trigeminal nerve at the foramen ovale.

November 24

1759

Tobias Smollett, "Smelfungus," is convicted of libel.

b.1885

Russell Morse Wilder, medical authority on diabetes and nutrition; he was the first to describe the symptoms of hypoglycemia.

d.1897

George Henry Horn, American physician and entomologist.

November 25

1783

Antonio Scarpa (1752–1832), Italian surgeon and anatomist, gives his first lecture at Pavia.

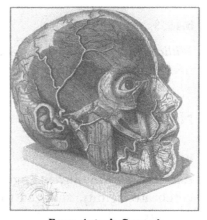

From **Antonio Scarpa's** *Practical Observations of the Diseases of the Eye*

1884

The first removal of a cerebral tumor is accomplished by British neurosurgeon *Rickman Godlee* (1849–1925).

b.1900

Gaetano Martino, Italian neurophysiologist, statesman, politician, and educator.

November 26

d.1686

Niels Stensen, Danish anatomist and physiologist.

b.1832

Mary Walker, American physician, woman's rights advocate; nurse, surgeon, and spy for Union Army in Civil War (Congressional bronze medal); trousered New York newswoman; invented the return postcard for registered mail; founder of "Adamless Eden," a colony for women only.

b.1853

William Murrell, English physician who (with *Lauder Brunton*) pioneered the use of nitroglycerin in treating angina pectoris.

November 27

b.1857

Charles Scott Sherrington, English neurophysiologist. He studied decerebrate rigidity, analyzed the stretch reflex, and described the "reciprocal innervation" of muscle groups. He introduced such terms as neuron, synapse, nociceptor, interoceptor, exteroceptor, teleceptor, proprioceptor, and proprioceptive.

November 28

1811

Decree by **Napoleon I** ending the School of Salerno. Its beginnings were lost in the legends of the ninth century, it was in Italy a "civitas Hyppocratica," an "Urbe Graeca." The *Regimen Sanitatis Salerni* was translated by John Harington:

> Use three Physicians still: first doctor Quiet,
> Next Doctor Merry-man, and Doctor Dyet...
> Joy, Temperance and repose,
> Slam the door on the Doctor's nose...

By the time of the Renaissance, **Petrarch** records its decline, "*In quo Salernum videbis. Fuisse hic medicinae fontem fama est, sed nihil, quod non senis exarescet.* Here Salerno may be seen, which formerly possessed fame as the source of the sciences, but now presents the decrepitude of dotage."

b.1873

Roberts Bartholow (1831–1904), American physician, the first to apply electrodes to the human cortex to demonstrate contralateral muscular contractions. This 1874 study was controversial because the experimental subject, who had given her consent, was a feeble-minded servant girl dying of a malignant purulent scalp ulcer.

1891

Dr. Thomas Neil Cream, wholesale poisoner, writes to *Dr. William Broadbent* to blackmail him regarding a case of strychnine poisoning. This letter would lead to *Cream's* capture and conviction.

November 29

b.1825

Jean Martin Charcot, French neurologist and gerontologist; he wrote the first descriptions of hysteria and was an influential teacher of

Sigmund Freud. He was a cosmopolitan personality, the founder of the *Archives of Neurology* and had many eponyms as credit to his clinical brilliance. "In the last analysis, we see only what we are ready to see, what we have been taught to see."

b.1902

Carlo Levi, Italian physician, author, painter, and anti-Fascist political writer. "The custom of prescribing some medicine for every illness, even when it is not necessary, is equivalent to magic, anyhow, especially when the prescription is written, as it once was, in Latin or in indecipherable handwriting. Most prescriptions would be just as effective if they were not taken to the druggist, but were simply hung on a string around the patient's neck like an abracadabra." *Christ Stopped at Eboli.*

1944

Alfred Blalock helps *Helen Taussig* demonstrate her proposed "blue baby" operation at the Johns Hopkins Hospital.

November 30

d.1603

William Gilbert, English physician and naturalist. In his *De Magnete* (1600) and other works, he wrote on magnetism, electricity, and scientific method.

> *Gilbert shall live till load-stones cease to draw*
> *Or British fleets the boundless ocean awe.*
>
> *—Dryden*

d.1694

Marcello Malpighi, Italian anatomist, naturalist, and botanist. The Singapore holly, *Malpighia coccigera*, is named in his honor.

1901

Viennese physician *Alfred Fröhlich* (1871–1953) reports a new syndrome of obesity, sexual infantilism, and hair loss associated with a hypothalamic tumor.

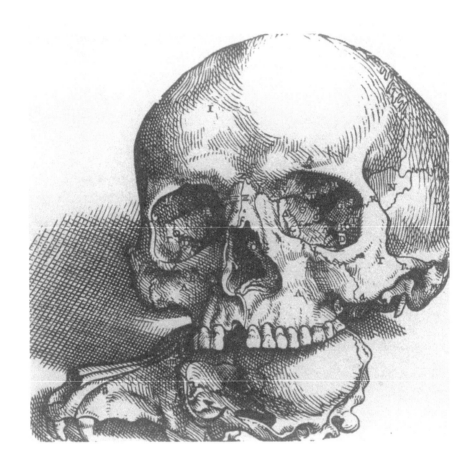

December

In a drear-nighted December.

—John Keats

December 1

Feast of *Saint Eligius*, bishop.

1712

Patrick Blair (c.1665–1728) is elected a Fellow of the Royal Society on the basis of his dissection of an elephant. In 1717 he wrote one of the earliest descriptions of pyloric stenosis.

b.1789

William Carr Lane, American physician, first mayor of St. Louis.

b.1802

Dominic John Corrigan, Irish physician who first described the "jerking" or "water hammer" pulse of aortic insufficiency that is named after him and noted that cardiac hypertrophy was an effect and not a cause of the disorder.

December 2

b.1885

George Minot, American hematologist who was awarded the 1949 Nobel Prize for his work in the use of liver to treat pernicious anemia.

1952

Childbirth is publicly televised.

1982

First mechanical heart is successfully transplanted into a patient, *Barney Clark*, by *Dr. William Jarvik*.

December 3

d.1910

Mary Baker Eddy, the Founder of Christian Science.

1967

Dr. Christiaan Barnard, at Groote Schuur Hospital in Cape Town, South Africa, performs the first successful heart transplant from *Denise Darvali* to *Louis Washkansky*.

December 4

1580

The Inquisition sentences *Doctor Garcia da Orta*, physician and professor of logic in India and Goa. He had written a pioneer scientific work on oriental medicinal plants and drugs, *Coloquios dos Simples & Drogas he causas medicinais da India* (1563) for which *Luis de Camões* had provided the introductory ode. He died in 1568.

1857

American naval surgeon *Edward Robinson Squibb* resigns to set up a reliable source of supply for drugs that the military will need for the coming war between the states.

1983

Jaime Lusinchi, a pediatrician, is elected president of Venezuela.

December 5

d.1624

Gaspard Bauhin, Swiss physician and botanist; *Bauhin's valve* is the ileocecal valve. He named the potato, *solanum tuberosum*. His brother *Jean* was the Father of Swiss botany, and the plant genus *Bauhinia* honors both brothers.

d.1771

Giovanni Morgagni, Italian physician and founder of pathologic anatomy.

December 6

1537

Vesalius is elected to the chair of surgery and anatomy at Padua.

1617

There is founded a new City Guild called the "**Masters, Wardens, and Society of the Art and Mystery of the Apothecaries of the City of London.**" For the decade prior to this the apothecaries had their privileges as a section of the powerful Grocers' Company.

d.1843

Abraham Colles, Irish surgeon.

1982

The execution of convicted murderer Charles Brooks becomes the first case of capital punishment by lethal injection.

December 7

1846

Thomas Bevill Peacock (1812–1882) describes the congenital association of pulmonic stenosis, dextroposition of the aorta, hypertrophy of the right ventricle, and ventricular septal defect—what would later become known as *tetralogy of Fallot*.

d.1847

Robert Liston, Scottish surgeon, inventor of *Liston's bone forceps*; he had a reputation for speed in pre-anesthetic surgery and was the first British surgeon to use ether.

Amputation of a leg

1905

Edouard Zirm, German surgeon, performs the first successful corneal transplant on *Alois Glogar.*

1967

Adrian Kantrowitz performs the first American heart transplant.

December 8

1617

First meeting of *Vincent de Paul's* Sisters of Charity.

b.1728

Johann Georg von Zimmermann, Swiss physician and philosopher.

b.1775

William Withering treats his first case of dropsy with foxglove (digitalis), as recounted in his *An Account of the Foxglove and Some of Its Medical Uses* (1785).

b.1797

Edmund Ravenal, American physician and planter, pioneer conchologist. He was the model for William Legrand in *Poe's* "The Gold Bug."

December 9

b.1748

Claude Louis Barthollet, French chemist who researched ammonia and chlorine compounds. He accompanied *Napoleon* to Egypt in 1798.

1784

In his will *Samuel Johnson* leaves to the physicians who cared for him without taking any fees "each a book at their election, to keep as a token of remembrance." When in his last illness *Dr. Johnson* was asked what physician he had sent for, *"Dr. Heberden,"* he replied, "ultimus romanorum, the last of our learned physicians." His classical learning led to frequent epithets for "virtuous and faithful *Heberden*" (*Cowper*) such as the eighteenth century *Hippocrates,* the English *Celsus.*

d.1884

Theodatus Garlick, American surgeon and sculptor who experimented with photography and was a pioneer in pisciculture (artificial trout-breeding). He published a classic paper on "Hybridization of Fish."

December 10

d.1198

Abû'l-Walîd Muhammad ibn Rischd (Averroes), Spanish physician and philosopher, governor of Andalusia, teacher of *Maimonides.* He wrote *Kitab-al-Kullyyat (Colliget).* He was persecuted and imprisoned for heresy: "Sit anima mea cum philosophis—let my soul be with the philosophers." *Galen, Rhazes, Avicenna,* and *Averroes* were reputed to be magicians and tricksters, and all passed into legend in the Middle Ages. Popular tales attribute to them various sorts of diagnosis and detection, with a *Sherlock Holmes* type resolution of problems of identity.

Avicenna

b.1804

Joseph Skoda, a Czech physician who modified and completed *Laennec's* work on auscultation and percussion.

1975

First successful graft of a totally severed male reproductive organ reported by Filipino surgeon *Ernesto A. Palanca.*

December 11

b.1803

Hector Berlioz. The medical career of budding musical composer *Berlioz* was aborted when the French government closed his medical school. "Become a doctor! Study anatomy! Dissect! Witness horrible operations instead of throwing myself heart and soul into the glorious art of music!....so I went on with my anatomy course, feeling no enthusiasm, but stoically resigned." *Memoirs,* 1870

b.1843

Robert Koch, German bacteriologist.

1844

Dr. Gardner Q. Colton administers laughing gas (nitrous oxide) to *Dr. Horace Wells,* and *Dr. John Mankey Riggs* then painlessly extracts a tooth.

1890

Emil Behring publishes an article documenting the serum therapy (antitoxin) and prophylaxis for diphtheria.

d.1948

Robert Stephen Briffault, English surgeon, anthropologist, and novelist. He practiced medicine in New Zealand; his most famous works are *The Mothers* (1927) and *Les Troubadours et le Sentiment romanesque* (1945).

December 12

b.1731

Erasmus Darwin, an English physician, poet, and philosopher. The influence of his *Zoonomia* (1794–1796) on the development of the theory of evolution was denied by his grandson *Charles*.

d.1777

Albrecht von Haller, Swiss physiologist, botanist, anatomist, and poet.

b.1889

George Washington Corner, American anatomist and embryologist who identified progesterone.

December 13

Feast of *Saint Lucy,* patron saint of ophthalmia and other diseases of the eye.

d.1204

Moses Maimonides, Jewish physician and philosopher.

1977

The World Health Organization reports the eradication of lethal smallpox.

December 14

b.1797

Emil Huschke, German anatomist remembered in eponymous structures of the eye, ear, and nose.

d.1916

Thomas Barr, Scottish otolaryngological surgeon who wrote the pioneer studies "Effects upon the hearing of those who work amid noisy surroundings," "Investigation into the hearing of school children," and "Giddiness and staggering in ear disease."

December 15

1748

John Mitchell (1680–1768), Virginia physician, botanist, and cartographer, is selected a fellow of the Royal Society. *Mitchella repens,* the checkerberry, is named for him.

d.1773

John Clayton, American physician and botanist. Spring beauty is named *Claytonia virginia* after him.

1820

The *Pharmacopoeia* of the United States is first published.

b.1859

Lazaro Ludovico Zamenhof, Russian-Polish Jewish ophthalmologist. He invented Esperanto and wrote *Linguo Internacia* (1887).

b.1878

Hans Carossa, Bavarian physician and autobiographical poet.

d.1924

Friedrich Trendelenburg, German physician and surgeon who has named for him a sign, a symptom, a gait, a test, a cannula, a position, and several operations.

December 16

d.1858

Richard Bright, English physician. *Bright's disease* is nephritis (a kidney disorder).

d.1940

Marie Eugene François Thomas Dubois, Dutch physician and paleontologist who discovered *Pithecanthropus erectus* or Java man.

d.1965

W. Somerset Maugham, English physician, novelist, and dramatist.

December 17

1776

From Bethlehem Hospital *Dr. William Shippen* writes sarcastically to his brother-in-law *Richard Henry Lee* about the American army,

"I wish you would introduce a new step into your army. I am sure they are perfect in the back step by this time."

b.1787

Johannes Evangelista Purkinje, Bohemian anatomist and physiologist. Cerebellar cells and a network of endocardial fibers are named after him; he pioneered in the study of subjective visual phenomena, described the first classification of fingerprints and was the first to prepare sections with a microtome. He also translated *Goethe* and *Schiller* into Czech.

b.1838

George Edward Post, American botanist and medical missionary to Syria. His *The Flora of Syria, Palestine and Sinai* described more than 120 new species and almost 300 new varieties of plants. *Postia lanuginosa* is named for him.

d.1881

Isaac Israel Hayes, American physician and Arctic explorer, author of *Open Polar Sea* (1867).

d.1887

Arthur Farre, English obstetrician, gynecologist, and microscopist. *Farre's white line* marks the attachment of the peritoneum to the ovary.

December 18

d.1892

Richard Owen, English surgeon and anatomist, paleontologist, zoologist, and museum conservator. He wrote *Odontography* (1840–1845); *Owen's line* lies across the dentine. He discovered the trichinosis parasite in human muscle.

d.1917

Josef Brudzinski, Polish physician who described several signs of meningitis.

December 19

d.1815

B.S. Barton, American physician, scientist, and botanist. A small herb in the gentian family, *Bartonia virginia*, is named for him.

b.1819

St. Julien Ravenal, American physician, Confederate surgeon, and agricultural chemist. He designed the torpedo cigarboat, the "Little David," which had an encounter with "Old Ironsides." He developed ammoniated and acid fertilizers and discovered that leguminous plants could restore worn-out soil.

1846

Dentist *James Robinson* performs the first dental extraction under anesthesia in Britain on *Miss Lonsdale* while ether is administered by *Dr. Francis Boott. Miss Lonsdale* experienced a "heavenly dream."

d.1947

Suicide of the English medico-legal expert and forensic pathologist, *Bernard Spilsbury*.

December 20

d.1590

Ambroise Paré, French surgeon, *un bon vieillard*. "Thursday, December 20, 1590, *Ambroise Paré*, surgeon to the king, died in his home in Paris, at the age of eighty. He was a learned man, foremost in his art, who, despite the time, spoke freely for peace and for the public welfare, which made him as much loved by good men as hated and feared by the wicked—which latter far outnumber the others." (*Pierre de L'Estoile, Memoires*) *Paré* introduced clean wounds, ligatures, and podalic version; he wrote *The Apologie and Treatise, Containing the Voyages made into Divers Places* (1585) *"Je le pensai, Dieu le guarist."* "God and nature doe sometime such things which seems to Physitions and Chirurgions to bee impossible."

1932

The first antibiotic, Prontosil, is discovered by **Gerhard Domagk** of I.G. Farben.

December 21

1846

Robert Liston (who could amputate a leg in 150 seconds) removes the putrescent thigh of *James Churchill* in the first European operation performed under anesthesia. Pre-operation: "We are going to try a Yankee dodge today, gentlemen, for making men insensible." Post: "This Yankee dodge, gentlemen, beats mesmerism hollow."

d.1967

Louis Washkansky, first human heart transplant subject.

December 22

d.1828

William Hyde Wollaston, English physician and physicist. He discovered that renal calculi consist of calcium phosphate, magnesium ammonium phosphate, calcium oxalate, uric acid, or their mixture. A method for making platinum malleable enabled him to retire from medicine. He discovered palladium, rhodium, and the Wollaston lens; he was a pioneer crystallographer who noted the identity of friction and current. The mineral Wollastonite is named in his honor as is the Wollaston Medal for research in mineralogy.

1836

Medical student **Charles Wadham Wyndham Penruddock** commits serious assault on his examiner **Mr. Thomas Hardy,** a surgeon, with intent to maim and disable him.

b.1872

Camille Guérin, French physician who with *Albert Calmette* introduced BCG, a vaccine for tuberculosis.

d.1902

Baron Richard von Krafft-Ebing, German psychiatrist who studied sexual abnormalities and wrote the classic *Psychopathia Sexualis* (1886).

December 23

b.1620

Johann Jakob Wepfer, Swiss physician and anatomist who was the first to inject blood vessels with colored liquids for the purpose of demonstration. His *Observationes anatomiae ex cadaveribus eorum quos sustulit apoplexia* (1658) revealed the etiological association between cerebral hemorrhage and apoplexy.

b.1804

Charles Augustin Saint-Beuve (1804–1869). His three years of medical studies were considered to have influenced his style of French literacy criticism.

b.1819

Carl Siegmund Franz Credé, the German gynecologist who developed a method of expressing the placenta, introduced silver nitrate to prevent ophthalmia neonatorum, and invented a premature incubator.

December 24

d.1298

Theodoric of Lucca (Teodorico Borgognoni, Theodoricus Cerviensis), Italian physician, *Bishop of Bitonto,* then of Cervica. A Dominican priest whose father was a surgeon to the Crusaders, he was a forerunner of antisepsis and anesthesia. He followed *Aristotle* (and the Chinese) and used urine to clean wounds; he also employed soporific sponges drenched with narcotics such as opium and mandragora.

b.1745

Benjamin Rush, physician, signer of the Declaration of Independence, "the American Sydenham." He was forced to resign as Surgeon General of the Military Department when his excellent penmanship revealed him as the author of a letter recommending the replacement of *George Washington.* He established the first free dispensary in America, was president of the country's first antislavery society, founded Dickinson college, served as Treasurer of the US Mint, and fought for penal and mental hygiene reforms.

Benjamin Rush

But he was a "drastic, idiotic phlebotomist" who seemed to believe, like *Gil Blas' Dr. Sangrado,* that blood was unnecessary to life. He was satirized by *William Cobbett* in *Peter Porcupine:* "He is a man born to be useful to society. (New York paper) And so is a mosquito, a horse-leech, a ferret, a pole cat, a weazel: For these are all bleeders, and understand their business full as well as *Doctor Rush* does his." *Rush* successfully sued *Cobbett* for libel and forced him to return to England.

b.1754

George Crabbe, English parish curate, surgeon and apothecary, poet with a grimly realistic portrayal of village life.

> *Anon a Figure enters, quaintly neat,*
> *All pride and business, bustle and conceit;*
> *With looks unalter'd by these scenes of woe,*
> *With speed that, entering, speaks his haste to go,*
> *He bids the gazing throng around him fly,*
> *And carries Fate and physic in his eye:*
> *A potent quack, long versed in human ills,*
> *Who first insults the victim whom he kills;*
> *Whose murd'rous hand a drowsy Bench protect,*
> *And whose most tender mercy is neglect.*

219

1890

In a letter **Anton Chekhov** refers to the inhibiting effect of an endo-toxin on the growth of malignant tumors.

December 25

b.1709

Julien Offray de La Mettrie, French physician, army surgeon, philosopher, and student of **Hermann Boerhaave.** A mechanist, his *Natural History of the Soul* got him expelled from France, and his *L'Homme Machine* got him expelled from Holland. He died while personal physician to **Frederick the Great of Prussia,** after treating himself for indigestion. "Everything gives way to the great art of the healer. The doctor is the one philosopher who deserves well of his country...The mere sight of him restores our calm....and breeds fresh hope."

1809

Dr. Ephraim McDowell, pioneer abdominal surgeon, removes an ovarian cyst from *Jane Todd Crawford* ("cousin apothecary" to *Mary Todd*) whom he had diagnosed on December 13, 1809—"A daring man and a courageous woman coming together to settle a problem." He was the mythic "doctor on horseback," and legend has it that while the operation was being performed in Danville, Kentucky, a noose was slung over a tree for Ms. Crawford's murderer.

b.1821

Clara Barton, American Civil War nurse who founded the American Red Cross.

1891

The first child receives diphtheria antitoxin in the Bergmann Clinic in Berlin.

1914

Edward C. Kendall isolates crystalline thyroxine.

December 26

b.1653

Johann Conrad Peyer, German nobleman, professor of logic, rhetoric, and medicine. In his 1677 *De glandulis intestinorum* he described the intestinal lymphatic patches or nodules named for him *(Peyer's patches).*

Edward C. Kendall

d.1780

John Fothergill, English physician and botanist, Quaker philanthropist. He wrote an *Account of the Putrid Sore Throat. Fothergill's diseases* are trigeminal neuralgia, diphtheria, and "sick headache" or migraine. With *Benjamin Franklin* he drew up a plan of reconciliation with the American colonies. When he made house calls to the poor, he frequently left more money than he took.

December 27

Feast of *Saint Fabiola* (d.399), a wealthy Christian who opened the first Roman institution for the care of the sick. A Roman Senator of the time, *Pammachius,* joined her in starting a hospital at Porta where she herself cared for the poor.

b.1822

Louis Pasteur, French chemist and bacteriologist.

December 28

b.1872

Pío Baroja y Nessi, Basque physician and novelist. His 22 volume cycle *Memorias de un Hombre de Acción* (1913–1935) is impersonal, pessimistic, chillingly intellectual, and deals with the rebels and pariahs of the world.

1895

Wilhelm Konrad Roentgen presents a paper before the Würzburg Physico-Medical Society, "A New Kind of Rays."

1895

The first clinical X-ray of a gunshot wound is performed by *Franz Exner* of Vienna.

The New Roentgen Photography: Look Pleasant Please."

December 29

d.1689

Thomas Sydenham, English physician. "Every merely philosophical hypothesis should be set aside, and the manifest and natural phenomena, however minute, should be noted with the utmost exactness.... By these steps and helps it was that the father of physic, the great Hippocrates, came to excel, his theory being no more than an exact description or view of nature. He found that nature alone often terminates diseases, and works a cure with a few simple medicines, and often enough with no medicines at all."

1750

Drs. John Williams and *Parker Bennet* of Kingston, Jamaica duel to the death in a dispute about the nature of "bilious fever." *Williams* had served as surgeon on slave ships in Africa and the West Indies and correctly distinguished yellow fever from malaria.

b.1863

Wilhelm His, Jr., Swiss anatomist who described the atrioventricular bundles of fibers named for him.

d.1919

William Osler, Canadian physician, founder and patron saint of Johns Hopkins Hospital. "Varicose veins are the result of an improper selection of grandparents."

d.1919

Egerton Yorrick Davis, mysterious physician, author of a number of disreputable, if not infamous, contributions to the medical literature.

December 30

1522

Publication of the *Isagogae* of *Jacopo Berengario da Carpi. Benvenuto Cellini* said that *Berengario* spent six months in Rome treating many patients with salves and fumigations (he was among the first to use mercurial ointment for syphilis); after he departed all his patients became worse and the populace threatened to kill him should he ever return.

1537

"I have also sent you an Almanack for the ensuing Year," *François Rabelais* in a letter.

d.1644

Jan Baptista van Helmont, Belgian physician, chemist, and Capuchin friar; he was a precursor of the iatrochemical school of medicine and the father of biochemistry. As a mystic he was given to apocalyptic visions that got him denounced to the Inquisition; he was acquitted two years after his death on the grounds of his life of piety. "*Helmont* was pious, learned, and famous, a sworn enemy to *Galen* and *Aristotle*. The sick never languished long under his care, being always killed or cured in two or three days." (**Lobkowiz**) His *De Magnetica vulnerum curatione* (1621) anticipated **Mesmer.**

b.1907

Lovelace W. Randolph II, NASA director of space medicine who selected and trained the first American astronauts.

December 31

b.1514

Andreas Vesalius of Brussels, anatomist author of *De Fabrica Humani Corporis* (1543). This date for his birth comes from a horoscope cast by the physician–mathematician *Jerome Cardan.*

1664

"But I am at great loss to know whether (my good health) be my hare's foote, or taking every morning of a pill of turpentine, or my having left off the wearing of a gowne." *Pepys' Diary.*

Medieval dissection

b.1699

Hermann Boerhaave, the Dutch Hippocrates. He set up organic chemistry as the basis of medical physiology and clinical observation as the preferred mode of medical teaching. The sweat glands are named for him. The sealed only copy of his self-published *The Onliest and Deepest Secrets of the Medical Art* sold for $20,000 at auction. The total contents of the otherwise blank book was the sentence: "Keep your head cool, your feet warm, and you'll make the best doctor poor."

b.1816

William Withey Gull, English physician who with *Allbutt* identified arteriosclerosis as a major cause of cardiovascular disease. *Gull's disease* is myxedema with atrophy of the thyroid gland. *Gull* was the master of the trenchant witticism: "The road to a clinic goes through the pathologic museum and not through the shop of the apothecary." "Savages explain, scientists investigate." A 1988 television movie identified him as "Jack the Ripper."

General References

Amos W: *The Originals: An A–Z of Fiction's Real-Life Characters*, Little, Brown, Boston, 1985.

Bailey H, Bishop WJ: *Notable Names in Medicine and Surgery* (3rd ed), Lewis, London, 1959.

Barnhart CL, Halsey WD: *The New Century Cyclopedia of Names* (3 vols) Prentice-Hall, Englewood Cliffs, NJ, 1954.

Boatner MM III: *Encyclopedia of the American Revolution*, McKay, New York, 1966.

Bordley J III, Harvey AM: *Two Centuries of American Medicine 1776-1976*, Saunders, Philadelphia, 1976.

Boussel P, Bonnemain H, Bové FJ: *History of Pharmacy and Pharmaceutical Industry*, Asklepios Press, Lausanne, 1983.

Castiglioni A: *A History of Medicine*, Knopf, New York, 1947.

Cuppy W: *How to Get from January to December*, Holt, New York, 1951.

Debus AG (ed): *World Who's Who in Science*, Marquis-Who's Who, Chicago, 1968.

Dobson J: *Anatomical Eponyms*, Baillière, Tindall & Cox, London, 1946.

Fort GF: *Medical Economy During the Middle Ages*, Bouton, New York, 1883.

Frewin A: *The Book of Days*, Collins, London, 1979.

Haggard HW: *Mystery, Magic, and Medicine*, Doubleday, Doran, Garden City, 1933.

Hendrickson R: *The Dictionary of Eponyms*, Dorset Press, 1988.

Johnson A (ed): *Dictionary of American Biography* (21 vol), Scribner, New York, 1946.

Jones NT: *A Book of Days for the Literary Year*, Thames & Hudson, New York, 1984.

Kelly EC: *Encyclopedia of Medical Sources*, Williams & Wilkins, Baltimore, 1948.

Kelly HA: *Some American Medical Botanists*, Southworth, Troy, New York, 1914.

Kenin R, Wintle J (eds): *The Dictionary of Biographical Quotation of British and American Subjects*, Knopf, New York, 1978.

Kronenberger L (ed): *Atlantic Brief Lives*, Little, Brown, Boston, 1971.

Leake CD: *Some Founders of Physiology*, Washington DC, 1956.

Long ER: *A History of Pathology*, Williams & Wilkins Co., Baltimore, 1928.

McGrew RE: *Encyclopedia of Medical History*, MacMillan, London, 1985.

McHenry LC: *Garrison's History of Neurology*, Thomas, Springfield, 1969.

McManus JFA: *The Fundamental Ideas of Medicine*, Thomas, Springfield, 1963.

Osler W: *The Evolution of Modern Medicine*, Yale Univ. Press, New Haven, 1921.

Parry LA: *Some Famous Medical Trials*, Scribners, New York, 1928.

Robertson P: *The Book of Firsts*, Bramhall House, New York, 1982.

Schmidt JE: *Medical Discoveries*, Thomas, Springfield, 1959.

Selwyn-Brown A: *The Physician Throughout the Ages* (2 vols), Capehart-Brown Co, New York, 1928.

Shosteck R: *Flowers and Plants: An International Lexicon with Biographical Notes*, Quadrangle, New York, 1974.

Silvette H: *The Doctor on the Stage*, University of Tennessee Press, Knoxville, 1967.

Stedman's Medical Dictionary, Williams & Wilkins Co, Baltimore, 1961.

Still GF: *The History of Paediatrics*, Oxford University Press, London, 1931.

Strauss MB (ed): *Familiar Medical Quotations*, Little, Brown, Boston, 1960.

Thorwald J: *The Century of the Surgeon*, Pantheon, New York, 1957.

Turner ES: *The Astonishing History of the Medical Profession*, Ballantine, New York, 1961.

World Almanac Dictionary of Dates, World Almanac Publications, New York, 1982.

Skulls of human and dog, from the *De Humani Corporis Fabrica* of Vesalius (1543); fig. 67, p. 124 in C. Singer, *A Short History of Anatomy from the Greeks to Harvey*, New York, Dover, 1957. Courtesy of the publisher.

The Temples and Cult of Asculapius, p. 35, *Great Moments in Medicine: The stories and paintings in the series A History of Medicine in Pictures* by Parke, Davis, Stories by G.A. Bender, paintings by R.A. Thom, Detroit, Northwood Institute Press, 1966. Reprinted by permission of Parke-Davis, a division of Warner-Lambert Co.

John Hunter (1728–1793), Plate 6; E.R. Long, *Selected Readings in Pathology*, Springfield: Charles C Thomas, 1961. Courtesy of the publisher.

Sir James Paget (1814–1899), from a portrait by George Richmond, 1867; p. 295 in R.H. Major, *Classic Descriptions of Disease*, Springfield: Charles C Thomas, 1945. Courtesy of the publisher.

Albert Schweitzer, Plate 26 (Arthur William Heintzelman), from C. Zigrosser, *Medicine and the Artist*, New York: Dover, 1970. Courtesy of the publisher and the Philadelphia Museum of Art.

St. Anthony and a victim of his disease (ergotism) p. 218, H.W. Haggard, *Devils, Drugs, and Doctors*, New York: Harper & Row, 1929.

The circle of Willis. Figure 1a in Thomas Willis, *Cerebri anatome: cui accessit nervorum description et usus*, Londini, J. Flesher, 1664. The engraving was made by Sir Christopher Wren. Figure 22, p. 61, in L. C. McHenry, Jr., *Garrison's History of Neurology*, Springfifield: Charles C Thomas. Courtesy of the publisher.

Thomas Vicary, Plate VIII, facing p. 67, Sir D'Arcy Power, *Selected Writings 1877–1930*, New York: Augustus M. Kelley, Publishers, 1970. With permission.

Hippocrates, p. 47 B.N. Engravings; P. Boussel, H. Bonnemain, F.J. Bové *History of Pharmacy and Pharmaceutical Industry*, Paris: Asklepios Press, 1983. Phot. Bibl. Nat. Paris.

Dr. and Mme. Carrel with Charles Lindbergh, Saint-Gildas, summer, 1936; facing p. 162, AM Lindbergh, *The Flower and the Nettle: Diaries and Letters 1936–1939*. New York: Harcourt Brace Jovanovich, 1976. Courtesy of Anne Morrow Lindbergh.

Florence Nightingale, Courtesy of the Harvey Cushing/John Jay Whitney Medical Library at Yale University.

Wilhelm Conrad Röntgen, Frontispiece, A.R. Bleich, *The Story of X-Rays from Röntgen to Isotopes*, Dover, New York, 1960. Courtesy of the publisher.

St. Apollonia, the patron saint of people plagued by toothaches, is at the same time the patroness of dentists. Meister von Messkirch. Albertina, Vienna , G. deFrancesco, "Saints in Medicine", *Ciba Symposia* 1: 108 (1939).

Silas Weir Mitchell. Figure 131, p. 328, L.C. McHenry, *Garrison's History of Neurology*, Springfield: Charles C Thomas 1969. Courtesy of the publisher.

Merrick, the 'Elephant-man" (after Treves), figure 258, p. 828, G. M. Gould, W. L. Pyle, *Anomalies and Curiosities of Medicine*, Saunders, 1986; reprinted by The Julian Press, 1956.

Claude Bernard (1813–1878) aged 53 (Courtesy of the Bibliotheque de l'Académie Nationale de Medicine, Paris), Plate 68. J.F. Fulton, L.G. Wilson, *Selected Read-*

ings in the History of Physiology, Springfield: Charles C Thomas, 1966. Courtesy of the publisher.

Charles H. Best and Sir Frederick Banting with dog no. 394, the first to be kept alive by insulin (1921). (Reproduced with kind permission of Dr. Best), Plate 90 J.F. Fulton, L.G. Wilson, *Selected Readings in the History of Physiology,* Springfield: Charles C Thomas, 1966. Courtesy of the publisher

The Wounded Man from the works of Paré, p. 16, Eaton Laboratories *Plasters, Pledges and Poultices: Wound Dressings through the Ages,* 1971. Reprinted by permission of Norwich Eaton Pharmaceuticals Inc.

William Cowper (1666–1709), From a portrait by Closterman, p. 340, R.H. Major, *Classic Descriptions of Disease,* Springfield: Charles C Thomas, 1945. Courtesy of the publisher.

Galen, Prince of Physicians next to Hippocrates (From Paré's *Surgery*); p. 339 H.W. Haggard, *Devils, Drugs and Doctors,* New York: Harper & Row, 1929.

Accouchement by a bull (after a 1647 engraving) Fig. 19, p. 133. G.M. Gould, W.L. Pyle, *Anomalies and Curiosities of Medicine,* Saunders, 1896; reprinted The Julian Press, 1956.

Bicephalic and hermaphroditic monster (after Paré) Fig. 26, p. 165. G.M. Gould, W.L. Pyle, *Anomalies and Curiosities of Medicine,* Saunders, 1896; reprinted The Julian Press, 1956.

Sir William Gowers. Figure 126, p. 313. L.C. McHenry, *Garrison's History of Neurology,* Springfield: Charles C Thomas, 1969. Courtesy of the publisher.

Robert James Graves (1795–1853). From the *Medical History of the Meath Hospital,* by L.H. Ormshy, Dublin, Fannon Company, 41 Grafton St., 1892. Kindness of George Blumer. From p. 279 in R. H. Major, *Classic Descriptions of Disease,* Springfield: Charles C Thomas, 1945. Courtesy of the publisher.

Descartes' figure of the heart showing the chorda tendinae; plate 58 from J. F. Fulton, L. G. Wilson, *Selected Readings in the History of Physiology,* Springfield: Charles C Thomas, 1966. Courtesy of the publisher.

Plate XII, William Hunter, *The Anatomy of the Human Gravid Uterus Exhibited in Figures,* London, 1774.

William Harvey, Plate 18 (Richard Gaywood), in C. Zigrosser, *Medicine and the Artist,* New York: Dover, 1970. Courtesy of the publisher and the Philadelphia Museum of Art.

Francois Rabelais, Plate 22 (Baltazar Moncornet), in C. Zigrosser, *Medicine and the Artist,* New York: Dover, 1970. Courtesy of the publisher and the Philadelphia Museum of Art.

John Browne, Frontispiece, *Myographia Nova,* London, 1697 reprinted New York: Editions Medicina Rara.

Marshall Hall (1790–1857), Plate 62. J.F. Fulton, L.G. Wilson, *Selected Readings in the History of Physiology,* Springfield: Charles C Thomas, 1966. With permission.

Leonardo da Vinci, 1452–1518, by himself, from a pastel in the Royal Library, Turin, Plate XIII, facing p. 94; C. Singer, *A Short History of Anatomy* from the Greeks to Harvey, New York: Dover, 1957. Courtesy of the publisher.

Carl Ludwig (1816–1895) Drawing by L. Knaus, 27 Aug. 1867. Courtesy of the National Library of Medicine, Plate 73, J.F. Fulton, L.G. Wilson, *Selected Readings in the History of Physiology*, Springfield: Charles C Thomas, 1966. With permission.

William Beaumont at the age of 66 (From an old daguerreotype) Plate 35. J.F. Fulton, L.G. Wilson, *Selected Readings in the History of Physiology*, Springfield: Charles C Thomas, 1966. Courtesy of the publisher.

Jean Fernel (1497–1558), From his *Therapeutices Universalis* (Frankfort, 1581) p. 647, R.H. Major, *Classic Descriptions of Disease*, Springfield: Charles C Thomas, 1945. Courtesy of the publisher.

Apollo cuts the Greek god Aesculapius out of the womb of his mother Koronia (Steingrueben Archive, Stuttgart) p. 201, Figure 33. Jürgen Thorwald, *The Century of the Surgeon*, New York: Pantheon, 1957.

Bones of the hand, from H. Gray, *Anatomy, Descriptive and Surgical.* Drawings by H.V. Carter, London: John W. Parker and Son, West Strand, 1858, p. 110, Plate 86.

Nude male, from the *Epitome* of Vesalius (1543); fig. 100, p. 188, in C. Singer, *A Short History of Anatomy from the Greeks to Harvey*, New York: Dover, 1957. Courtesy of the publisher.

Santiago Ramón y Cajal (1852–1934), Plate 24, E.R. Long *Selected Readings in Pathology*, Springfield: Charles C Thomas 1961. Courtesy of the publisher.

Dr. Jared Potter Kirtland (Courtesy of Case–Western Reserve School of Medicine) p. 910, G.E. Gifford, Jr. "Dr. Jared Kirtland and His Warbler", *New England Journal of Medicine* (1972) vol. 287. Reprinted by permission.

Treating (? causing) wry neck. Plate facing p. 98, vol I, Nicolas Andry, *Orthopaedia*, London, 1743.

Edward Jenner (1749–1823) Plate XL, f.p. 144 in E.R. Long, *A History of Pathology* New York: Dover, 1965. Courtesy of the publisher.

St. Erasmus's intestines being wound up on a windlass. 1450–1460 Staatsbibliothek Munchen. G. deFrancesco, "Saints in Medicine," *Ciba Symposia* 1: 115 (1939).

Nude female, from the *Epitome* of Vesalius (1543); fig. 100, p. 188, in C. Singer, *A Short History of Anatomy from the Greeks to Harvey*, New York: Dover, 1957. Courtesy of the publisher.

The Sir William Osler bookplate, 1919, Max Brodel, *Hopkins Medical News* November 1979, p. 4. Courtesy of Ranice W. Crosby, The Brödel Archives, Art as Applied to Medicine, The Johns Hopkins School of Medicine.

Surgical Instruments from Pompeii (by the courtesy of Prof. K. Sudhoff) Figure 7, facing p. 22. in C. Singer, *From Magic to Science: Essays on the Scientific Twilight*, New York: Dover, 1958. Courtesy of the publisher.

Theophrastus Paracelsus. Drawing by Hirschvogel, 1566 edn. of Paracelsus' writings. Frontispiece, F. Hartmann, *Paracelsus: Life and Prophecies* Blauvelt NY: Rudolf Steiner Publications, 1973. Reprinted by permission of Garber Communications.

Plate I from *Myographia Nova*, London, 1697, reprinted New York: Editions Medina Rara.

Pasteur in his laboratory, Plate 20 (Timothy Cole),in C. Zigrosser, *Medicine and the Artist*, New York: Dover, 1970. Courtesy of the publisher and the Philadelphia Museum of Art.

Larrey's "Flying Ambulance," p. 14, Eaton Laboratories *Plasters, Pledges and Poultices: Wound Dressings through the Ages,* 1971. Reprinted by permission of Norwich Eaton Pharmaceuticals Inc.

The fashionable Dr. Marat, physician to the aristocracy. G. Weissmann, "Marat on Sabbatical", *Hospital Practice*, vol. 10, #7, July 1975, p. 105. Courtesy of the Bettman Archive.

Skeleton of the "Irish giant" in the Royal College of Surgeons, London. John Hunter bribed the undertakers almost 500 pounds to obtain the body against the expressed wishes of the giant. Fig. 157, p. 330 in G.M. Gould, W.L. Pyle, *Anomalies and Curiosities of Medicine,* Saunders, 1896; reprinted by The Julian Press, 1956.

Caesarian section after a seventeenth century woodcut, p. 214, Figure 38 (Steingrueben Archive, Stuttgart) Jürgen Thorwald, *The Century of the Surgeon,* New York: Pantheon, 1957.

Guy de Chauliac (1300–1368). R. H. Major, *Classic Descriptions of Disease*, Springfield: Charles C Thomas, 1945, p. 78.

St. Pantaleon, physician and martyr, is called upon in cases of headache. Detail from the frescoes in the Church of San Domenico a Taggia near Imperia, Ligur, Piedmont. School, 15th century. G. deFrancesco, "Saints in Medicine", *Ciba Symposia* 1: 112 (1939)

Muscles of the back, plate 143, p. 219, Henry Gray, *Anatomy, Descriptive and Surgical* drawings by H. V. Carter, MD. London: John W. Parker and Son, 1858.

Medieval Hospitals: La Grand' Chambre de Povres, part of the Hotel-Dieu, Chartered August 4, 1443, first patient admitted January 1, 1452. From p. 69 in *Great Moments in Medicine: The stories and paintings in the series A History of Medicine in Pictures* by Parke, Davis, Stories by G.A. Bender, paintings by R.A. Thom, Detroit, Northwood Institute Press, 1966. Reprinted by permission of Parke-Davis.

Girolamo Fracastoro (1483–1553) Plate 1 in E.R. Long, *Selected Readings in Pathology*, Springfield: Charles C Thomas, 1961. Courtesy of the publisher.

Joseph Lister (1827–1912) p. 121, H.W. Haggard, *Mystery, Magic, and Medicine,* Garden City: Doubleday, Doran, 1933.

St. Roch, the most popular protector against the plague. Plague broadside by Michael Wolgemut, Nurmberg, ca. 1484. G. deFrancesco, "Saints in Medicine", *Ciba Symposia* 1: 110 (1939).

Oliver Wendell Holmes, (Courtesy of Medical Classics) p. 830. F.A. Willius, T.E. Keys, *Classics of Cardiology*, New York: Dover, 1961. Courtesy of the publisher.

Helmholtz: Physicist–Physician, inventor of the ophthalmoscope. From p. 229 in *Great Moments in Medicine: The stories and paintings in the series A History of Medicine in Pictures* by Parke, Davis. Stories by G.A. Bender, paintings by R.A. Thom, Detroit, Northwood Institute Press, 1966. Reprinted by permission of Parke-Davis, a division of Warner-Lambert Company.

231

First muscle tabula, from the *De Humani Corporis Fabrica* of Vesalius (1543); fig. 105, p. 193 in C. Singer, *A Short History of Anatomy from the Greeks to Harvey*, New York: Dover, 1957. Courtesy of the publisher.

Relative proportions of men and women, from Robert Knox, *Artistic Anatomy*, London: Henry Renshaw, 1852, p. 49.

Interior view of a modified Auvard incubator. Fig. 6, p. 69, in G.M. Gould, W.L. Pyle, *Anomalies and Curiosities of Medicine*, Saunders, 1896; reprinted The Julian Press, 1956.

Dante and his book from the picture by Domenico de Michelino, Florence, facing page 193 in P. Toynbee, *Dante Alighieri*, London: Methuen, 1900.

Cosmas and Damian, by Thomas de Leu, born in Flanders, active in Paris from 1560–1620. Kupferstichkabinett Berlin. J. Gerlitt, "Cosmas and Damian, the Patron Saints of Physicians," *Ciba Symposia* 1: 118 (1939).

François Magendie (1783–1855) Plate 61 in J.F. Fulton, L.G. Wilson, *Selected Readings in the History of Physiology*, Springfield: Charles C Thomas, 1966. Courtesy of the publisher.

Harvey Cushing (1869–1939), taken in his Baltimore laboratory *ca.* 1912. Plate 67 in J.F. Fulton, L.G. Wilson, *Selected Readings in the History of Physiology* Springfield: Charles C Thomas, 1966. Courtesy of the publisher.

Chart of veins (workshop of Titian) in the *De Humani Corporis Fabrica* of Vesalius (1543); plate 76 in C. Zigrosser, *Medicine and the Artist*, New York, Dover, 1970. Reproduced by permission of the publisher & the Philadelphia Museum of Art.

Nicholas Tulp (1593–1674), Engraving by L. Visscher. The frontispiece to *Observationes medicae* (Leyden, 1716); p. 140.; R.H. Major, *Classic Descriptions of Disease;* Springfield: Charles C Thomas, 1945. Courtesy of the publisher.

Andreas Vesalius, Plate 17 (Jan Stephen von Calcar), in C. Zigrosser, *Medicine and the Artist*, New York: Dover, 1970. Courtesy of the publisher and the Philadelphia Museum of Art.

St. Luke, a patron saint of physicians, mitigates the sufferings of the dying. From a matriculation volume in the university library of Basel, 1484. G. deFrancesco, "Saints in Medicine" *Ciba Symposia* 1: 102 (1939).

Sir Thomas Browne, Plate 23 (Robert White), in C. Zigrosser, *Medicine and the Artist*, New York: Dover, 1970. Courtesy of the publisher and the Philadelphia Museum of Art.

Ambroise Paré, Plate 16 (German School, XVI century), in C. Zigrosser, *Medicine and the Artist*, New York: Dover, 1970. Courtesy of the publisher and the Philadelphia Museum of Art.

Antony van Leeuwenhoek (1632–1723) from his *Epistolae ad Societatem Regiam Anglicam*, 1719. Plate 12 in B. J.F. Fulton, L.G. Wilson, *Selected Readings in the History of Physiology*, Springfield: Charles C Thomas, 1966. With permission.

Henry Koplik (1858–1927). From p. 170. H. Ellis, *Bailey and Bishop's Notable Names in Medicine and Surgery*, London: H.K. Lewis, 1959. Courtesy of Chapman and Hall.

Second skeleton in the *De Humani Corporis Fabrica* of Vesalius (1543); figure 103, p. 191 in C. Singer, *A Short History of Anatomy from the Greeks to Harvey,* Dover, 1957. Courtesy of the publisher.

Dissection by Woman Medical Student, 1870. Engraving; p. 72 (National Library of Medicine) A. Novotny, C. Smith, *Images of Healing,* New York: Macmillan, 1980.

The plague doctor—the costume used by physicians during the plague in Marseilles in 1720. The snout was filled with spices presumed to purify the inhaled air. An engraving by Paulus Furst; p. 209 in H.W. Haggard, *Devils, Drugs, and Doctors,* New York: Harper & Row, 1929.

Marie Curie, N.U.M.L. Portrait Collection, in G. Marks, W.K. Beatty, *Women in White,* New York: Scribner's, 1972.

Plate I from Antonio Scarpa, *Practical Observations on the Diseases of the Eye,* London, 1806.

Plate prepared by Casserius and published in Adrian Spigelius, *De formato Foetu* (Padua) with a dated dedicaation of 1626); figure 96, p. 169, Charles Singer, *A Short History of Anatomy and Physiology from the Greeks to Harvey.* New York: Dover, 1957. Courtesy of the publisher.

Mid-Calf Amputation of the Left Leg in the 18th Century (Courtesy of the Boston Medical Library). From *A General System of Surgery,* L. Heister (1683–1758) (Third English language edition, London, 1748) p. 1044, J.W. Estes, "The Practice of Medicine in 18th Century Massachusetts", *New England Journal of Medicine* (1981) vol.. 305. Reprinted by permission of the *New England Journal of Medicine.*

Avicenna, a frequently encountered painting. On page facing p. 52; in H.C. Krueger, *Avicenna's Poem on Medicine,* Springfield: Charles C Thomas, 1963. Courtesy of the publisher.

Benjamin Rush (1745–1813), Portrait by Sully. From the engraving by Edwin, in *American Medical Biography,* by J. Thacher (Boston, 1828); p. 228 in R. H. Major, *Classic Descriptions of Disease,* Springfield: Charles C Thomas, 1945. Courtesy of the publisher.

Edward C. Kendall. Plate 92 in J.F. Fulton, L.G. Wilson, *Selected Readings in the History of Physiology* Springfield: Charles C Thomas, 1966. Courtesy of the publisher.

"The New Roentgen Photography: Look Pleasant, Please" A cartoon from *Life,* February 1896 (courtesy the Armed Forces Institute of Pathology) Fig. 3, facing p. 10. A.R. Bleich, *The Story of X-Rays from Röntgen to Isotopes.* Dover: New York, 1960. Courtesy of the publisher.

Mondino directing a dissection; from a Fifteenth-century woodcut. Figure 40, p. 94. C. Singer, *From Magic to Science: Essays on the Scientific Twilight,* New York: Dover, 1958. Courtesy of the publisher.

Index